DATE DUE

DE 19 97			
AP 28 99			
JE 2 00			
JE 8 05			

DEMCO 38-296

About Island Press

Island Press is the only nonprofit organization in the United States whose principal purpose is the publication of books on environmental issues and natural resource management. We provide solutions-oriented information to professionals, public officials, business and community leaders, and concerned citizens who are shaping responses to environmental problems.

In 1994, Island Press celebrated its tenth anniversary as the leading provider of timely and practical books that take a multidisciplinary approach to critical environmental concerns. Our growing list of titles reflects our commitment to bringing the best of an expanding body of literature to the environmental community throughout North America and the world.

Support for Island Press is provided by Apple Computer, Inc., The Bullitt Foundation, The Geraldine R. Dodge Foundation, The Energy Foundation, The Ford Foundation, The W. Alton Jones Foundation, The Lyndhurst Foundation, The John D. and Catherine T. MacArthur Foundation, The Andrew W. Mellon Foundation, The Joyce Mertz-Gilmore Foundation, The National Fish and Wildlife Foundation, The Pew Charitable Trusts, The Pew Global Stewardship Initiative, The Rockefeller Philanthropic Collaborative, Inc., and individual donors.

About The Nature Conservancy

The mission of The Nature Conservancy is to preserve plants, animals, and natural communities that represent the diversity of life on Earth by protecting the lands and waters they need to survive.

To date the Conservancy and its members have been responsible for the protection of more than 9 million acres in 50 states and Canada. It has helped like-minded partner organizations to preserve millions of acres in Latin America, the Caribbean, the Pacific, and Asia. Although some Conservancy-acquired areas are transferred for management to other conservation groups, both public and private, the Conservancy owns more than 1,400 preserves—the largest private system of nature sanctuaries in the world.

Beyond the Ark

Beyond the Ark

Tools for an Ecosystem Approach to Conservation

W. William Weeks

Foreword by Bruce Babbitt

Island Press
Washington, D.C./ Covelo, California

ISLAND PRESS is a trademark of The Center for Resource Economics.

Library of Congress Cataloging-in-Publication Data

Weeks, W. William.
    Beyond the Ark : tools for an ecosystem approach to conservation / W. William Weeks.
        p.   cm.
    Includes bibliographical references and index.
    ISBN 1–55963–392–1 (cloth). — ISBN 1–55963–393–X (pbk.)
    1. Ecosystem management.   2. Nature conservation.   3. Nature Conservancy (U.S.)   I. Title.
    QH75.W43   1997
    333.7'2—dc20   96–23724
                                                                          CIP

Printed on recycled, acid-free paper

Manufactured in the United States of America
10 9 8 7 6 5 4 3 2 1

Contents

PART III
Ecosystem Conservation in Context   125

Preface and Acknowledgments

I work for The Nature Conservancy. The organization began a decade ago to apply goodly parts of its substantial experience and resources to the problems of ecosystem conservation, and it has been working systematically at it for half of that time. Many other conservationists in both private and public sectors have, in recent years, developed an interest in the same topic. No one has it figured out. In fact, the Conservancy's former director of science used to say that to be really good at ecosystem conservation you have to be an expert on everything. That is a tough test to pass, and the Conservancy's people and programs haven't passed it. But while mastery has so far proved elusive, the pursuit has revealed enough that it seems worthwhile to write some of it down. I know that my having done so will not make an expert of anyone, but I think that it could help the Conservancy and others meet more effectively the increasingly complex challenges of working at conservation in this fast-developing world.

When I first began to think of writing this book, I thought that the real learning to be shared was contained in the approach to ecosystem planning that occupies part II. By the time I got around to preparing a short outline, I was envisioning the book as an exercise only a bit more reflective than a manual. I expected to set forth the planning discipline that some Conservancy colleagues and I developed to help our project teams consider more thoughtfully the choices they had to make in pursuing the conservation of fairly large ecological units of the landscape. I hoped to make the Conservancy's approach accessible to a variety of people undertaking and interested in conservation challenges.

I think the book will be a useful guide, and I have tried hard to make it accessible; it isn't a manual. Almost from the first words I wrote, I realized that I had been trained, formally and informally, to look at conservation and conservation problems in a particular way, the Conservancy's way. The planning discipline that got me started writing proceeds from a number of assumptions that are embedded in the Conservancy's way of doing business, and for an audience not very familiar with the Conservancy, a lot of those assumptions need to be made explicit. The planning methodology was indeed designed to help conservationists make choices, but the choices that need discussing include not only those that arise within a conservation project but also some choices that precede, and others that supersede a conservation project.

So rather than writing a manual, I have paused at many places to make explicit the assumptions that come as second nature to Conservancy people and, occasionally, to discuss them analytically. I have also tried to provide some conservation and social context for the planning advice given. I think that those changes in approach have made the book more meaningful, and I hope they have made it more interesting as well.

I have ventured opinions upon some subjects that I would have chosen not to address if I were writing or speaking for the Conservancy. The Conservancy's public statements about the importance of conservation tend to be consistent with the organization's approach to conservation: rational and pragmatic. But Conservancy people tend to have deep feelings for the land and for nature. The personal reflections on conservation are, therefore, true to a part of the Conservancy's spirit. On those few occasions that I have chosen to offer comment on social institutions, by contrast, the Conservancy would surely and rightly want to keep its distance. All in all, however, the book is more about the Conservancy's approach to conservation than anything else.

I will qualify the previous statement by noting that the Conservancy is full of smart and strong-willed people. Unanimity of opinion is almost unheard of, and consensus, difficult enough. Further, the organization, like all lively things, is constantly changing. There is some disagreement among Conservancy people about many of the points I make, and that includes three ideas that are central to the presentation: whether it really is good strategy to look beyond nature preserves for optimum protection of biodiversity; whether land or biodiversity conservationists ought to use substantial resources dealing with human, social, and economic questions; and whether we really need to conserve, in some way, very large parts of what I have called the seminatural landscape. Whatever the nature of the organization's doubts and disagreements, it has been moving rapidly into work that implies affirmative answers to all three questions. Nearly all of the examples I present are from Conservancy experience.

Other people and other organizations are working to develop approaches similar to those presented here. But while I have tried to retain a little part of the journalist's objective eye and voice in this writing, I have not tried to be a journalist. I am a conservationist. I have firm opinions about conservation that were formed as I worked with Conservancy people on some of the projects discussed. I know the conservation work I describe better than any journalist could afford to. I can discuss it with a fair degree of confidence. I can't say that about the work of, for example, Ecotrust or the Sonoran Institute, though from what I do know, their efforts seem to be promising and well worth learning about.

The book is divided into three parts. The first is an introduction to conservation planning, to conservation, to The Nature Conservancy, and to the relatively new and rather popular notion of ecosystem management. The second presents the conservation planning discipline that was the original impetus for the book. In the third, some tangents and implications of the approach generally advocated in the book are considered. I have also included in part III, as a conclusion to the book, a personal reflection about conservation.

*   *   *

The methodology for planning an ecosystem approach to conservation was developed over some time and with a lot of input from the project teams that were trying to put it to use. Greg Low of Reston, Virginia, provided particularly important leadership and insight, and he, David Harrison of Boulder, Colorado, and George Fenwick of Washington, D.C., deserve special credit (if there is credit to be distributed) for many of the useful features of the discipline. I deserve blame for any lack of clarity in the manner in which it is set forth here, and for all mistakes in the text. I have also, many times, taken basic points of the planning discipline the Conservancy developed and expanded on them because it seemed to me that telling the story or answering questions that arose in the telling demanded it; for those departures, neither the Conservancy nor the above-named group bears any responsibility, and because it would be tedious to cite each of those instances, my creative colleagues can safely deny responsibility for anything I have written.

The influence of Conservancy thinking on my view of conservation will be clear to readers who know the Conservancy; a dozen years of sometimes contentious adherence to its distinctive conservation worldview has not been without its effect. Even so, aside from the planning methodology itself, the commentary and conclusions, influenced by twenty years of reading and thinking and talking about conservation, are a synthesis that is as original as it can be under the circumstances. Certainly the Conservancy cannot be held responsible; indeed, I have, in this forum, considered and consciously decided to comment on issues I would have chosen not to address if I were writing for the Conservancy.

In that general regard, a word about usage seems appropriate. I have often used the word "we" as if we are all together trying to solve a conservation problem, and I have also used it to refer either to the community of conservationists or to society in general—to attitudes and actions for which I think there is collective responsibility. I have tried to avoid using it to refer to The Nature Conservancy, even though that has become a habit. When I began to work for the Conservancy in 1982, I felt awkward saying "we" because it implied some personal responsibility for accomplishments that preceded my time. After a while, I felt instead as if my efforts had joined those of others in the current of a great river, an ongoing collective enterprise—and that "we" was the correct pronoun for Conservancy work. Having already used the word to mean three different things in this book, however, I thought a fourth was a little much, and I have usually referred to the Conservancy as would an observer rather than a participant.

To my colleagues in the Conservancy, whose wrong turns seem to be as well documented here as their successes, know that I consider you leaders and heroes. Conservation requires new pathways, and you don't find them by walking the well-trod way. Your successes—ours—are the reason this book has been written. Our mistakes may serve to make the way clear for those who will follow. One couldn't hope for a more inspiring job description.

*   *   *

There are other important acknowledgments to make as well, and though it seems pretentious to thank as many people as I'd like to thank for an achievement as modest as this, the recognition and appreciation I offer is genuine, as well, inevitably, as incomplete.

First and most of all, I want to thank Mary, my wife, for her knowledge and love of nature and her unique insights as a reviewer.

I also want to warmly thank The Nature Conservancy's board of governors and its exceptionally able president, John Sawhill, for permitting me to spend half my time over ten months writing this book.

It is important, too, in my judgment, to recognize the people who made TNC an organization that could do the work I describe and who instilled the values that make it effective. Because the list would otherwise be so long, I have confined myself to citing only three by name: Pat Noonan, Greg Low, and Bob Jenkins.

Literally millions have made and are making contributions of time, talent, advice, and money to the Conservancy, and there are hundreds whose contributions are so distinguished that they deserve special mention in a book that implicitly celebrates the capabilities of the Conservancy. But when I have tried to do justice to that task, I have found it impossible. I have known scores of very special Conservancy people myself but will mention only Sally Reahard and Dan Efroymson, who have been exceptionally important friends, advisors, and colleagues during my years of work with the Conservancy.

I want to thank Barbara Dean, who worked hard to get me to address a certain inclination to laziness in both storytelling and explanation; she has made me be about as clear as I am going to be, and I must say she was both kindly and firm in approaching all of the considerable editorial challenges my work presented. I have come to admire her perspective. Her colleague, Chuck Savitt, correctly insisted that TNC needed to get some of its thinking into the literature and encouraged me, in the absence of another alternative, to do some writing myself. Pat Harris did a remarkably attentive copyedit, truly improving the text. The superfluous articles and sentence fragments that remain in the text are not there because she missed them.

Bob Jenkins generously provided a thoughtful and extensive critique of the first full draft text. I haven't followed all his advice, but as usual, I have always learned from it. Greg Low and Brian Kahn also reviewed the text for me, and their comments and questions similarly made the book a far better product. Mike Coda, Trisha Klein, and James Workman, too, provided valuable assistance. And I am truly grateful, as readers of this book will be, to Bruce Babbitt for providing a foreword that does all that anyone could ask of this device.

My mother, Mary, gently instilled in me a sense of wonder and the capacity for a personal relationship with nature, as well, I suppose, as the confidence to try something like this.

That is where the book has come from. Joseph, John, and Anna Weeks are my children, and they are where we are going. My hope is placed with joy in them, and the work I describe is, more than anything else, done for them.

Foreword

There has been a recent tendency for authors to write about conservation struggles in America from the point of view of an armchair spectator. Often the observers have never played or coached the conservation game, witnessed the months of training, studied the playbook, or uncovered the tense personal struggles within the team itself. They see their jobs as simply a matter of explaining, with dismissive confidence or cheerful complacency, who won and why, then moving on to what the team should have done in the first place.

Beyond the Ark breaks out of that simplistic formula. Bill Weeks writes not only as an individual who has worked and fought and gotten seasoned in the land management arena while keeping his savvy and vision intact; he writes as someone ready and willing to stay there, tackling complex projects and seeing them through from murky beginnings, to incremental transformations until finally they are set on a course to a hopeful future.

That element of hope for the future is critical. Through his eyes we come to realize the complexity and the pressures behind even the smallest day-to-day decisions. With critical analysis, he brings the reader to points in each issue where despair or indifference would be an understandable escape. Time after time, he proves there is a fine line between a critical or skeptical mind and a mind that has grown merely cynical. Weeks never crosses that line.

The cynic sits back and points out "Not only will that conservation approach fail, it will lead to degradation of the landscape and create a political backlash. Give it up." The skeptic, like Weeks, replies, "You're right, it may not make sense over the long term. But let's sit in the uncomfortable chairs of this conference room with the buzzing lights and the stale coffee until we hammer out something that does."

Admittedly, it is often a haphazard, messy process where no single party gets 100 percent of its desired outcome. It is a process that results in compromise—a dirty word to many environmentalists and polluters alike. But it is a democratic process in the oldest sense of the word. The satisfaction comes from the fact that it builds a flexible coalition of individuals who support a common model for lasting conservation. It gives them a blueprint for managing their landscapes of complexity.

One major value of Weeks' book is that he gives these blueprints—and the consensus-building stories of the people who designed them—to urban, suburban, and rural communities that are anxious to preserve their local heritage, the familiar wildlife, trees, rocks, plants, and watersheds that give them a sense of place.

A greater value, I think, is how he carefully weaves the human condition into that sense of place, giving an added strength and texture to nature's rich and productive tapestry. Some environmentalists have focused on the serious conservation issues surrounding cattle ranching. However, they have not come to grips with the fact that, if cattle ranching becomes unprofitable, often larger ranches will be subdivided into numerous smaller parcels of land, which can have even more devastating effects on biological diversity. The goal is an approach where the economic forces are brought into the decision-making process, and enhance the ecology through, in Weeks' term, "compatible management."

Finally, a compelling aspect of *Beyond the Ark* is the author's willingness to advocate bold economic strategies that address key problems where they occur. As we increase the scope, we increase the stakes and the pressures on decision makers. But as he shows in the discussion of the Virginia Eastern Shore Sustainable Development Corporation, Americans are proving up to the challenge. The strategies may not work in every instance, but the willingness to grapple with them—skeptically, critically, but far from cynically— shows that the democratic process remains a vibrant, if messy, force for change.

There is nothing naive about this book, nothing wooly minded or old fashioned; nor does it offer any abstract, utopian silver bullets that will solve our problems for us. The author cannot afford such luxuries. If the author can care enough about our local and familiar watersheds to take a hard-headed, hands-on approach to restoring them, so can the rest of us.

Bruce Babbitt

The Conservancy, Conservation, and Ecosystem Management: The Context of the Planning Discipline

Figure 1-1. The Gray Ranch, Animas, New Mexico. © Ron Geatz

CHAPTER 1

Introduction

Early in December of 1989 The Nature Conservancy's New Mexico state director called me at TNC's home office to say that we had a decision to make about the Gray Ranch. We knew the place fairly well; the New Mexico program staff and volunteers had been looking at the ranch on and off for a couple of years and had concluded that it ought to be a Conservancy priority. Others in the conservation community were equally familiar with the ranch. The Wilderness Society, for example, saw the Gray as a model for a new set of additions to the National Wildlife Refuge System, these to be selected for biodiversity importance.

The ranch was controlled by a Mexico-based business concern. The owners didn't particularly care about the Gray; it is said they had acquired it as part of a package of assets that became theirs when they foreclosed on a loan made to finance a Denver high-rise. Preliminary discussions with the owners about purchasing the ranch had been terminated early in 1989, when there seemed to be no way to close the $5 million gap between our respective ideas of a fair sale price. Exasperated, finally, with the conservation attention and resulting traffic of officials the ranch had attracted, the owners closed the place to outsiders late in the spring of 1989.

The owners of the ranch reopened negotiations seven months later by informing us that they had an offer for the ranch in hand. They issued an ultimatum: beat that price—several million dollars lower than what they had been asking in the spring—or they would accept the new offer at the end of the week. We got in touch with the competing would-be buyer. He had no interest in our overtures regarding possible avenues of cooperation. We reviewed and revisited all the information we had about the ranch. We listened to lobbying efforts by some, including a couple of United States senators, who wanted us to buy it. On the other hand, Manuel Lujan, then Secretary of the Interior, let us know that he did not want the Gray as a new wildlife refuge. That dashed any illusion we had that protecting the Gray would be a simple matter of purchasing the land and recovering our costs over a few years with a smooth and uncontroversial transfer to the U.S. Fish and Wildlife Service.

There was more to our interest in the Gray than another deal and another refuge, anyway. The Conservancy had begun to wrestle, in 1989, with the intellectual and institutional issues associated with launching a systematic effort to protect a set of large conservation areas, because it was becoming increasingly apparent that the extensive system of mostly relatively small nature preserves we had already established was an insufficient response to a biodiversity crisis. We hoped, by establishing some large reserves, to protect more biodiversity more securely. We had also begun to believe that conserving and restoring characteristic ecological processes was essential to our biodiversity goals and that large conservation projects would often be required to sustain ecological processes.

While the Conservancy was, as I have said, preparing to launch an effort to establish some large conservation areas, our preparations were far from complete as we considered our response to the ultimatum on the Gray. At 325,000 minimally disturbed acres, the Gray sure seemed like the kind of place we had in mind for the new system. During the seven days the owners had given us to consider buying the Gray, we made the decision to leap ahead of the intellectual foundation we were developing for an ecosystem conservation initiative. We bought the ranch without knowing exactly what was there, how we would manage it, how we would raise the money needed, or exactly what the Gray Ranch would contribute to the system of larger conservation areas we were talking about establishing. We had enough scientific information to know that the place was important, but we were acting mostly on emotion. For conservationists, emotion is the easy part, and the Gray— big, scenic, diverse, unconquered, and full of the spirit of wilderness—called up powerful emotions.

Our emotional response to the Gray reinforced our conviction to act upon a business philosophy that a great conservation organization ought to take great risks for great places. Great conservation, though, requires a sound overlay of planning and strategy on a foundation of emotion. There wasn't time to develop such an overlay before we committed to the purchase of the Gray. Buying it was the right thing to do; indeed, it seems, kind of surprisingly, to have been the decision to acquire the Gray that catalyzed and sealed the Conservancy's commitment to establish the system of larger conservation areas that we had been discussing. But it has been a difficult project, more difficult than it would have been had we been able to make better plans for its management.

The difficulties began before we had even closed the deal. One New Mexico senator's aide wanted the Conservancy to endorse authorizing legislation that would clear the path for transfer of the entire ranch to the U.S. Fish and Wildlife Service. The other senator from the state was interested in partial transfer to the Service and retention of the rest in private ownership with conservation easements—anathema at the time to the aide representing the first senator. We became aware that the local community didn't much like the Conservancy as an owner but positively hated the idea that the U.S. Fish and Wildlife Service, which had done a study predicting 30,000 visitors a year, would take ownership. We wondered too about the Service's plans for the refuge and suggested some changes in the proposed legislation. We were interested in having the law set forth a clear mandate to manage the property with biodiversity values as a principal objective. Some experts on the U.S. Fish and Wildlife Service including a fellow from The Wilderness Society, hated the idea of including such directions in legislation. They didn't want Congress so closely directing the activity of the agency. Not that they always agreed with the Service; part of their job was to find instances in which the agency was serving its mission poorly. But the Society's representative didn't want Congress to get into the business of providing detailed prescriptions for refuge management.

We had confidence in the Service's regional director, and he had openly expressed his interest in the Gray as a biodiversity refuge. So, in spite of the disagreements we were having about the authorizing legislation, we worked to keep the possibility of transfer alive. The secretary of the interior's position had not changed from the time we reopened negotiations to buy the property; indeed, his views were so strongly held that he did an amazing thing during our negotiations with the owners of the ranch: he called them to tell them personally that the Service had made absolutely no commitment to repurchase the ranch from the Conservancy; he wanted to make certain, presumably, that we weren't telling them that the money we would use to pay them was going to be generated in such a transfer. His flirtation with tortious interference diminished our reluctance about trying to go over his head to the White House, and we were advised by an inside Washington player to meet with a White House aide. The result of the meeting was equivocal. Apparently we failed to emphasize the features of the ranch that were expected to get the aide excited about the place, and the White House simply let it be known that it would not oppose the proposed authorizing legislation, stopping short of offering affirmative support.

Faced with authorizing legislation we didn't much like, conflicting signals from the executive branch, community opposition to a sale to government, and an increasingly obvious shortfall in information about the ecology of the ranch, we did the first really measured thing we had done in the project to that time. We informed the senators that while we would not speak against the authorizing legislation, we would not endorse it either, and that before we made any decision about selling the ranch, we were going to take a year to conduct ecological studies and manage the ranch ourselves, to find out what the ranch was really all about and to make careful plans for its future.

One senator took the news in stride, but an aide for the other wasn't shy about expressing his outrage. He decided to pressure the Conservancy to change its mind and called a friend and political ally who had an impressive record of conservation accomplishment and who was, not coincidentally, chair of the Conservancy's New Mexico chapter. The resultant rebuke stung. A highly accomplished Conservancy volunteer leader was angrily accusing us of robbing the public of access to a dream. We were shaken, but we stood firm for slowing the process down a bit, at the least. Within a couple more days we were summoned to a meeting with another advocate of the refuge, a key staffer of the House Subcommittee on Interior Appropriations. He demanded to know what we were up to. My bland answer didn't satisfy him. He leaned forward and said, "You say you are going to study this place for a year, but what I want to see and don't see is the big wink. I don't see the commitment to get this thing into the hands of the Fish and Wildlife Service." I was amazed. He wasn't finished. "You all are up here a lot. You conduct a lot of business with this committee. You have to understand that I'm like a stamp collector, and I value my collections enormously. And my collection won't be complete until the Gray Ranch is part of it. Think about that while you're doing your studies on the ranch."

If we didn't have a plan for the ranch when we bought it, it was becoming clear that some other folks surely did. We took the year, even so, and took the time to think about what this very special place meant to conservation. It certainly could be a premier nature preserve. Some Conservancy people thought we ought to keep the place and manage it as such. It would become one of the organization's flagship places. Alternatively, it could be a fine wildlife refuge, and it might even be perceived as such a success that it would reinforce what some perceived as a healthy trend that was not secure within the Service—the purchases of refuges for reasons that had little or nothing to do with ducks. Finally, the ranch, we also found during that year, was a place of unusual importance in the community of cattle growers. The Gray had been the southernmost unit of the well-known Diamond A—a million-acre ranch assembled around the turn of the century by the Hearst family of publishing fame. Even in its current configuration, the Gray was a big and famous old-fashioned western cattle ranch and one of those places that an awful lot of people who watch the cattle business knew about.

We began to wonder whether the greatest conservation leverage the ranch could provide might lie in its status as a working cattle ranch on which biodiversity thrived. We had thereby begun an evolution in our thinking that would eventually lead to an ongoing effort to work with the owners, public and private, of a million acres in the Borderlands terrain around the Gray. We're still adjusting our thinking and our plans, but the conservation project, it seems clear, will have expanding and enduring importance. The people we angered by taking the planning fork in the road early in 1990 have mostly recovered, at least enough to work with us again. But if we had thought the project through better and earlier, if we had even been able to announce at the outset our intention to take a year to study the place and to develop a plan for its future, a lot of unhappiness and wasted energy might have been avoided.

We had, nonetheless, shown ourselves in the Gray Ranch project that there was great promise and important satisfaction in working on bigger conservation projects, and it appeared likely that we had the capacity for doing so. We had gained a glimmer of an understanding about a practicable way to conserve biodiversity on a large scale and about the role economic uses of land might play in meeting that challenge. We had also shown ourselves that we had better find a management system with more organization and predictive power than ad hoc decision making.

We have learned something about each of those themes since buying the Gray Ranch. This book sets forth much of that learning.

CHAPTER 2

Planning and Action

Conservation is an increasingly wide-spreading tree. The main stem, it seems to me, has always been sustainable use of natural resources, but there was an early branch devoted to protection of nature for nature's sake. Some branches emerging mostly from the main stem, such as pollution control, have grown wide and strong. That branch and others are intertwined with those of other trees—civil engineering, climatology, chemistry, law, and politics. There are new branches, too, growing toward other trees such as international trade and economic and racial justice.

If you start climbing the tree these days, it's easy to lose track of where you are among the tangle of branches. Part of our problem at the Gray Ranch was that we leaped right into the thick of the crown. The likelihood of reaching the place you want to be can be improved by standing back a little from the tree before you start and thinking about how best to get there. Or, to say it another way, to make a plan.

Conservation planning is (to recall a moment of hope that brightened the generally dismal series of recent presidential campaigns) the "central and organizing" theme of this book. A word about planning, therefore, seems appropriate at the outset.

I began my career in conservation as something between a skeptic and an unbeliever with respect to the proposition that planning was a good thing. It seemed to me that things changed too fast for planning more than a couple of weeks in advance to be useful. My perspective was flawed.

I was working as a state director for The Nature Conservancy, an organization that has made good use of planning. A lot of planning was being done and had been done around me. The organization had, for example, made a conscious decision that it could make the most progress toward conserving biodiversity in the United States by fostering the establishment of staffed units in each state. The Conservancy had thoughtfully considered and decided what it wanted these state programs and their directors to do, what support it would provide, what guidance it would give, how the work would be evaluated, and what quality controls it would impose. The Conservancy had an organization-wide plan that provided reason, context, and guidance to the work of its state directors.

Also, I had developed what would, by objective standards, be called my own plan for the state program. It was broad and it wasn't detailed, but I knew where I was going. I thought we could preserve the biodiversity of the state and build recognition and financial support for our work by substantially increasing the number and geographic distribution of land acquisition projects we completed. I thought the state's Division of Nature Preserves was a natural partner, and I wanted to build on the good relationships the organization already had with the DNP. Further, I wanted to attract to our cause the interest of an increasingly better connected group of business, philanthropic, and government leaders who could enlarge our capacity to do conservation within the state. I structured my decisions about what actions to take, what new projects to pursue, where to spend my time, and generally how to use the scarce resources that I had at my disposal in accordance with that broad outline of a plan.

After I had worked in the Conservancy for a few years, planning became second nature. We made annual plans, strategic plans, fund-raising plans, and project plans. If they do nothing else, reasonably good plans document understandings about priorities and make accountability possible. Planning can do considerably more. It can be an aid to orderly decision making and incisive thinking.

Dick Heckert, former chairman of DuPont, put the case for planning this way during a committee meeting of TNC's board of governors: "If you don't care where you're going, any road will take you there." Dick wasn't suggesting that the Conservancy didn't care. He meant that the "any road" approach won't work if you do care. The methodology described in this book was developed not only to help Conservancy project teams that were working on fairly large ecological units know where they were going but also to help them look over and compare a variety of roads that would take them there.

Having said those positive things about planning, it's worth examining in a little more detail the objection I used to have to planning. There was, in fact, more to it than the feeling that any plan I made would soon be made obsolete. I was also making a statement about resource allocation: I objected to planning as a substitute for action. I still have that objection. More often than some of us would care to admit, planning is displacement activity for the risks and rigors of work in the real world.

I would go so far, in fact, as to assert that we ought to be acting at the same time we are planning, always. If we aren't compelled to do so by the importance and the urgency of the challenge we have taken on, we aren't working on a sufficiently critical challenge.

A well-conceived plan will guide an active conservationist toward taking the right next step, but it won't do much more than that. It is actually *taking* that next step—acting—that makes possible that most important of opportunities, the opportunity to react. If the plan is particularly good, the proper reaction will have been anticipated and preparations for it made in advance. But the best of plans will eventually run thin on foresight and must then be revisited. Planning as seen in the Conservancy is an organic and an iterative process; that is, both plan and planning process must be revisited, must grow, develop, and change as the work unfolds.

Saying this does not mean that whatever result is achieved will be acceptable. The planned destination ought to be changed much more slowly and cautiously than the planned path for attaining it. And the accountability made possible by planning requires that even the failure to reach or the choice to bypass mileposts along the way be explained. To whom is the explanation owed? The people doing the planning and the work owe it to themselves first. Then perhaps to their supervisor, their investors, their funders, and others who review their work.

The Conservancy has made extensive, perhaps even excessive, use of processes for reviewing plans. It has invested most heavily in reviews of the first iteration of a plan because that is a natural time to stop, reflect, and take one more look, with knowledgeable outside help if possible, at the hypotheses that have been proposed and the decisions that have been made. The completion of the first iteration of a plan is an important time in the life of a conservation project, something like a commencement as that word is used in graduation ceremonies. It is a natural time to slow down for a few minutes of review and reflection. The data available for review is obviously richer at subsequent stages of a project. But getting fresh and critical input is important, and it is hard to make time and arrangements to do it. We are justified, therefore, in seizing and making the most of a review at the natural break point after the completion of the first iteration (not to say that we shouldn't push for periodic review in later stages of projects—we should).

Though it is almost always useful, it isn't always easy to find a reviewer who is sufficiently grounded in the goals and context of the conservation project, but sufficiently objective to provide that "fresh and critical input." If it is impractical to obtain review by a knowledgeable but less invested reviewer for either an original plan or a report of results and a new draft plan, I recommend a review of the plan by the planners themselves after a vacation, a weekend, or even a weekday afternoon of fly fishing. The idea is to put some distance between plan and planner, to reclaim some objectivity.

Whatever method of review is used, the objective is not, again, to confirm development of the final and immutable truth about how to proceed with the project. When the first few field office–generated plans for conservation of large and complicated natural systems were presented to reviewers from the Conservancy's home office, the purpose of the exercise was understood to be approval of the plans. It became clear early on that stamping a plan "approved" sent the wrong signal; it suggested that the planning process was complete. The review process was soon recast so as to make projects, not plans, the subject for approval. The point of that relatively subtle change was to emphasize that although the plan was a culmination of a process involving a lot of work and energy, it wasn't *the* culmination. An approved plan goes on the shelf as a finished product. In the worst case there is a psychological satisfaction that important work has been done. It hasn't. A step has been taken, and useful preparation for work has been done. But the real work is conservation, not planning, and even the planning work itself is not finished. Rather, one good iteration has been finished in a planning process that should motivate more work and produce many more iterations.

Because the plan is not a thing to be written in stone, and is not in fact the real work that needs to be done, we are back to the question of whether planning is really such an important thing after all. The answer is . . . it isn't. But the planning process, the thing that I choose to call a discipline, is. In most projects of substantial scope, there will be some steps that it is clear from the start ought to be taken, and after all, the surefootedness we need to be effective planners is quite often best obtained by walking. Before too long, though, we're bound to overtake the point made in slightly different words by Thomas Jefferson, and others before and since. As I learned the aphorism: "It ain't what we don't know hurts us so much; it's what we know for sure that just ain't so." Some of our early actions will, in retrospect, have been unnecessary or even counterproductive. And after those first steps have been taken, the next steps must be chosen from among an imposing number of choices, most of them wasteful if not directly harmful. It is in making the right choices after taking those first steps that a planning discipline is especially useful.

Or, to return to the metaphor of the wide-spreading tree of conservation, if you start climbing up the wrong stem, you may have a tough time crossing over to the right one, and the climbing is a rigorous enough task all by itself.

CHAPTER 3

The Nature Conservancy in the Time of the Flood

I have made considerable reference already to The Nature Conservancy. My feelings about nature come from a wide variety of experiences, but what I know about how to conserve nature comes directly or indirectly from my work with the Conservancy. It's good practice, when evaluating advice, to consider the source. A little direct description of the Conservancy is therefore in order.

Other than a book of "most of what follows is true" stories (*Good Dirt*, by David Morine), and a coffee table book, in which an interesting descriptive text by Noel Grove is easy to overlook among the Stephen Krasemann photographs (*Preserving Eden*, 1990), the last generally available book setting forth a Conservancy approach to conservation was *Building an Ark: Tools for the Preservation of Natural Diversity Through Land Protection*. It was written by the Conservancy's Phillip Hoose and published in 1981. The ark of Hoose's title is a system of "specific areas of land" protected because they "provide the biological requirements" of plants, animals, and natural communities native to the United States. The ark, the book explains, could be built with a set of (still extraordinarily useful, by the way) tools for choosing and conserving those specific areas of land.

If the prescribed ark wasn't being built in the middle of the flood then, the deep water has surely arrived now. That is, in part, the explanation for the development of the conservation methodology described in this book. While some things have remained the same, a lot has changed since the Conservancy was explicitly and implicitly described in *Building an Ark*. The critical drivers of change in the Conservancy are, interestingly, set forth in the introduction of *Building an Ark* as challenges that neither the book nor the Conservancy had yet addressed. Hoose gently laments the national focus of his presentation; since 1981, the Conservancy has developed very active international efforts. Hoose says the book doesn't address the protection of aquatic species; the Conservancy has since accepted that challenge. Finally, Hoose says that stewardship of conserved areas isn't addressed; it is the implications of that obligation, as much as anything else, that have moved the Conservancy away from relying solely on the gathering up for the ark of "specific areas," spots on the landscape, as a response to the flood.

A brief discussion of the Conservancy will update the portrait that emerged from *Building an Ark* and will, as well, begin to define the frame of reference from which the conservation opinions and experiences in this book have been set forth.

The Nature Conservancy was incorporated in 1951 for the purpose, broadly stated, of preserving natural areas. By the mid-1950s, it had begun to settle on land acquisition as the principal means of pursuing its purposes. The mission statement was sharpened in the 1970s, and since that time the Conservancy has addressed itself to the preservation of biodiversity. In its most recent iteration, the Conservancy's mission is "to preserve plants, animals, and natural communities that represent the diversity of life on earth by protecting the lands and waters they need to survive."

When it chose to focus on biodiversity protection, the Conservancy carried forward the approach that had characterized its first twenty years; that is, it continued to distinguish itself by taking direct action, usually the acquisition of land, to conserve the places it was concerned about. The Conservancy's accomplishments are considerable. Almost 10 million acres in the United States have been protected through Nature Conservancy efforts, and management efforts have been brought to bear on millions of additional acres of parkland in Latin America and the Caribbean through Conservancy assistance to partner organizations based in those regions. In recent years the Conservancy has also established conservation programs in a number of Pacific island nations and in Indonesia.

Of the ten million acres of land protected in the United States, a little less than a million acres is bound up in nature preserves owned and managed by The Nature Conservancy itself, and another two million acres have been leased or are managed by TNC. This set of privately owned natural areas is thought to be the largest such system in the world.

Almost four million acres are now protected after a Conservancy gift, sale, or assistance to various government entities, local, state, and federal, for inclusion in their systems of parks and protected areas. These transfers are, as a matter of policy, made at the Conservancy's cost or below. More than 1.5 million in additional acres of public land have been designated for enhanced conservation management through Nature Conservancy work.

Of the remaining acreage protected by Conservancy efforts, three quarters of a million is owned privately by entities other than TNC and is legally protected because development rights have been conveyed permanently to the Conservancy through instruments such as conservation easements. Lands comprising another 750,000 acres have been protected through a variety of Conservancy programs and are now held by a diverse group of owners, public and private. Many of the largest such tracts are owned by other conservation organizations and by universities.

A decade ago, The Nature Conservancy was a significant but smaller and less well known environmental organization. In the middle of the 1990s it is, by many measures, the largest conservation organization in the country and one of the largest in the world. It is supported by more than 800,000 members. (Both definition of membership and counting methodology vary in the nonprofit world, incidentally. The Conservancy includes in its membership count donors to conservation projects as well as those specifically contributing dues, and it claims both for thirteen months after a contribution. Those are comparatively conservative standards.)

The Conservancy's annual operating budgets have moved steadily upward toward $150 million, and its capital expenditures add more than $100 million to the organization's conservation impact. More than 1,500 scientists, businesspeople, lawyers, land managers, and support staff members work for the Conservancy, and they can be found in more than 200 business locations, including offices in every state.

The efforts of this staff are informed and guided by a network of biological inventories staffed by additional scientists, working mostly in state government. The institutional participants in this network, generally referred to as Natural Heritage Programs, gather biological and conservation information and log it into Conservancy-designed computerized databases. The enterprise is the embodiment of the vision of Robert Jenkins, who was the Conservancy's director of science from 1970 to 1993. The Heritage network does many things, but prominent among the results it achieves is the identification of places that are important for biodiversity conservation before destruction is imminent so that conservation action can be taken in an atmosphere of deliberation and reason, cost can be kept to a minimum, and resolutions in which all interested parties come away satisfied are likely.

The Conservancy supports and continually updates the software network members use to keep biodiversity data, and it maintains a central node in the network. There are participants in every state, in four Canadian provinces, in fourteen Latin American and Caribbean countries (outside the United States these biological inventories are usually known as Conservation Data Centers, or CDCs), and in a number of national parks and forests. The information gathered and organized by these units has as a critical foundation the work of many natural historians, past and present, and the network relies on cooperative relationships with the community of universities, museums, and botanical gardens. Heritage and CDC units have, in turn, added greatly to the scientific community's store of plant and animal records by their own fieldwork.

Heritage field records are electronically filed for access by either biological or geographic inquiry. Judgments about conservation priority are maintained along with other parts of the record. The databases are therefore useful not only for setting conservation priorities but also for making decisions about the location of planned development. Network participants handle thousands of inquiries annually from government, industry, and conservation organizations. When, for example, a road, a pipeline, or a power substation is being planned, the proposed route or location can be checked against Heritage records, and if the development would damage a population of a rare plant or a nesting habitat of a declining bird, the plans can be altered while it is still inexpensive to do so. Through that capability alone, an enormous amount of conservation good is done.

The efforts of the professionals working in and closely with the Conservancy are magnified by the work that thousands of volunteers contribute toward biodiversity goals in preserve management and natural area restoration, as well as in office administration. The Conservancy receives tens of thousands of hours of freely contributed help annually, some of which literally couldn't be bought. Among the corps of Conservancy volunteers there are pilots and mapmakers, students and professors, photographers and network television newspeople.

A critical element of both the Conservancy's growth and its way of doing business has been the establishment of state and local operating units. These units are guided and assisted by local civic leaders, conservationists, and scientists organized into chapter boards. Members of these boards serve without pay and are, in legal effect, advisory. However, they have real authority under Conservancy policy, and their service is central to the successful operation of programs. The Conservancy grants, and expects the combined local staff and volunteer programs to accept, substantial responsibility for planning and funding conservation efforts in and from their base of operations. That autonomy is understood to be accompanied by an obligation to pursue work that clearly furthers the organization's mission and to do so in a manner consistent with its stated values. The management of the Conservancy, substantially decentralized in that fashion, can be flexible enough to allow and, ideally, encourage experimentation and innovation precisely because it is very clear about mission and values. Conservancy people know the organization's mission and accept the values that will guide them in pursuit of that mission—not the methods, but the manner in which the mission will be pursued.

It would be a bad thing for the conservation movement if all environmental organizations shared all the Conservancy's values; different approaches are needed and different niches must be filled, just as in the natural world. But it is important to highlight several of TNC's values because they say a lot about the organization. The Conservancy says, of itself, that it works "through nonconfrontational means toward tangible and lasting results." Its orientation is toward solving problems rather than making judgments. Its approach tends to be pragmatic, though it is driven by highly idealistic goals. It says that its work requires "vision, resourcefulness," and an "entrepreneurial spirit and adaptability to change," yet it prides itself on and is notable for its adherence to a value of "continuity of purpose." Those values taken together create tension, and that constructed tension is critical to the Conservancy's strength.

The Conservancy's values statement includes a commitment to meeting a perceived responsibility to future generations and a statement of belief that, as individuals and collectively, the organization can make a difference. Yet it does not delude itself that it can accomplish its mission by its efforts alone. The statement of values also acknowledges the need, because of the scope and urgency of its mission, to involve "all sectors of society, public and private."

Finally, the Conservancy says, significantly, that the organization will use "the best scientific information available." Within that phrase is both a commitment to science and a commitment to action—that is, an acknowledgment that it is the best available science, not scientific certainty, that is required for action.

The manner in which the Conservancy works with government illustrates a number of its values. To the greatest degree possible, government officials are considered and treated as colleagues and partners. That inclination extends to the way the organization handles its differences with government. It seldom publicly criticizes government action or officials. Its style is to differ privately, in the presence of the responsible officials. To prepare for a meeting to discuss differences with government representatives, TNC typically makes a serious effort to develop a course of action the government could follow that would resolve the points of disagreement in a manner that satisfied the objectives of both sides.

When, for example, a draft management plan for a national forest is released for public comment and the "preferred alternative" for management fails to incorporate a commitment to protect a significant natural area, the Conservancy is unlikely to release a statement for the media criticizing the plan, the planners, and the forest supervisor. It is more likely to quietly prepare to submit comments on the plan by arranging a meeting with the supervisor and staff, bringing along good documentation of the relative importance of the area in question, trying to gain an understanding of the objective that motivated the planners to leave the critical area unprotected, and considering with the Forest Service team ways to achieve that objective without compromising the natural area.

That kind of approach does not always work. But it does give government officials the respect they are due, it does acknowledge the complexity of the interests they are balancing, and it does represent a way of expressing differences that any of us would appreciate were we in their shoes. It also preserves the relationship for communication on other, possibly even more important, differences tomorrow.

Partnerships of the kind TNC has with government have their limits, of course. Conservancy policy makes Conservancy cost the standard for transferring land to government agencies. For many years, the definition of the cost of land acquired by the Conservancy for the federal government included the cost of money—that is, interest accrued or forgone while Conservancy funds were bound up in conservation land that was acquired for transfer to a government agency. In recent years, cautious government officials have reinterpreted the applicable spending regulations (perhaps fearing criticism for developing too close a relationship with nonprofit conservation organizations), and cost has been redefined. The new cost formula ignores interest but includes a figure for the Conservancy's overhead costs. Nobody much likes paying somebody else's overhead. The Conservancy had historically been willing to absorb that cost in service of the mission and as a contribution in the spirit of the partnership. Interest, though, is another matter. It is passing strange to account for federal government transactions as if the money the Conservancy uses to buy important properties doesn't cost anything. One doesn't build or maintain a stable relationship by cultivating a costly fiction. Openness and candor must be the foundation of any working partnership, and the new cost formula chips away some of that foundation. That is dangerous, because stress from routine disagreement and minor irritation is encountered in the normal course of partnerships. When additional stress is introduced, the survival of the partnership can become an issue. There are, fortunately, also opportunities to strengthen the foundation. The Conservancy has contributed tens of millions of dollars in land value to various units of government over the years, and the government has managed most of that land quite well. That helps.

The Conservancy's willingness to transfer land at cost, forgoing the opportunity to resell to the government priority lands at whatever the market will bear, is a reflection of two organizational conclusions. First, it reflects a realization that the available resources are insufficient to adequately protect the nation's biodiversity and that those we have ought to be stretched to the furthest extent possible. Second, it reflects a desire to treat those in government who are involved in preserving biodiversity as true and valuable partners. There are conservation organizations that follow a different policy, buying properties they believe could be government priorities at the best price possible, selling them to government at the government-appraised value, and financing their operations largely on the margin. It is good that these organizations exist because there is a lot of work to be done in land conservation, much of it for purposes distinct from those of the Conservancy. The mutual willingness of government agencies and those organizations to do business on that basis indicates that there is a need for the service being provided. But the generous spirit that attends the Conservancy's at-cost policy somehow feels better and is, in the end, less vulnerable to criticism. These days, there is reason to be wary. There are increasingly vocal and active opponents of the government's—especially the federal government's—owning land as part of its role in conservation. That criticism has the potential to reduce government's conservation activity and make cooperation difficult. And while we might not save enough biodiversity to fulfill our duty to our heirs by working cooperatively, we're certain to fall short if we fail to cooperate.

Disparate efforts will fail not because they are themselves unimportant or insubstantial but because the conservation challenge is enormous. Size is relative. The Nature Conservancy was earlier described as the largest conservation organization in the United States. Does that make it large? The Conservancy's total assets exceed a billion dollars, though that figure includes the value of land that is held for nature preserve purposes and therefore neither would nor, in many cases, could be converted to cash. TNC isn't small even by the standards of the larger not-for-profit world, having in recent years been consistently ranked among the nation's twenty-five largest not-for-profit organizations, using annual contributions from private sources (individuals, corporations, and foundations) as a measure.

The Nature Conservancy has, moreover, raised enough capital for its work to establish a revolving fund of nearly $170 million for land acquisition. This capital, administered as the Land Preservation Fund, puts the organization in a strong position to compete when rural properties of the highest biodiversity interest become available for purchase. Further, when that fund has been fully subscribed through the land acquisition activities of the Conservancy's field offices, the organization's strong balance sheet has given it access to reasonably priced lines of credit.

That kind of capability defines the Conservancy as large even by comparison with the dreams of its founders, and in the swirl of numbers and expressions of amazement at what the organization has become, it is easy to forget what "large" can mean in a broader context. Compared with the institutions of the for-profit world, TNC is not large. Mobil, to pick an example because a report on its recently closed quarter happens to be close at hand, made half a billion dollars in *profit* during the *three months* covered by the news release.

The Conservancy, to bring this small point to a close, is of moderate size when compared with other enterprises. Even more important, it is tiny when considered next to its mission. In a good year, it might bring conservation action successfully to a close on a thousand acres of land per day. Three times that much land is converted to urban or suburban use in the United States on an average day; 40 billion tons of earth are moved every year.

Even more to the conservation point, the U.S. Fish and Wildlife Service recently estimated that 300,000 acres of wetland are drained or filled every year (subsequent reports suggest that the rate of loss may have slowed somewhat). The last really sizable tracts of tropical hardwood forest are being logged under authority of concessions such as those being considered by the government of Suriname at the time of this writing for 7 million acres of its land. It is estimated that worldwide, nearly 30,000 species are driven to extinction every year.

The size of the effort that the Conservancy and all others taking action on the land to protect species and natural communities quite clearly pales before the scope of the problem. The Conservancy's contribution toward achievement of its mission needs to grow significantly, and for that to happen the Conservancy will have to continue to grow significantly, even if it learns as well to do more with the resources it has.

One way it can grow, of course, is by attracting more contributions. Because TNC is a so-called public charity under the federal tax code, contributions of land, the bargain (below-market) portion of bargain sales of land, and qualifying interests in land (such as irrevocable transfers of development rights through conservation easements), as well as money and personal property, are deductible for federal income tax purposes to the fullest legal extent. The Conservancy has marketed the benefits of such deductions effectively and creatively, though the lower federal income tax rates of the late 1980s and 1990s have made the tax benefits of gifts a relatively less important factor in donor motivation. In recent years, simple personal commitment to the cause of conservation and, for some donors, the benefits of public recognition have been more central motivations. More and better tax incentives would, of course, help generate the increased support that the organization and the cause need.

With respect to existing sources of support that make the Conservancy's work possible, about 70 percent of contributions to the Conservancy come from individuals, about 15 percent from foundations, and about 10 percent from corporations. Government grants and contracts have, in the past few years, increased from being a negligible source of income to a substantial one (about 15 percent of budgeted income, considering all sources), with contracts for providing assistance in the protection of parks in Latin America leading the way.

Of corporate support, the founding father of the modern Nature Conservancy, Pat Noonan, used to say what Booker T. Washington is supposed to have said: "It may be tainted money, but 'taint enough." There is more to the question than that, of course, but for the Conservancy, there was and is wisdom in the fundamental point. That is, TNC is willing to recognize donors for the good they are doing through their gifts, will accept those gifts without reservation, and will provide wholehearted recognition. The Conservancy neither condemns nor even much assesses the broader environmental record of its donors.

This attitude has earned the Conservancy some occasional flak from the environmental community. But to those who have said the Conservancy is "selling out," I would say that integrity doesn't require that TNC confront at every opportunity those whose environmental record could be better. If honor invariably required confrontation, we'd quickly become pretty tiresome company; few of us know any organization or individual that it would not always be our duty to confront for some commission or omission. Indeed, we might carefully examine our own lives for environmental sin before deciding to cast the first stone. Selling out, properly considered, can only be declining to do something one otherwise would have done (or doing something one would not have done) in exchange for support that otherwise would not have been granted. The Conservancy is vigorous, activist, and competitive in pursuing the strategies it has set forth for preserving biodiversity, and it is vigilant in defending the rare species and communities on its preserves. It conducts processes that include both volunteers and staff to determine its generally pragmatic strategies. Instances in which a donor has tried, even by subtle means, to constrain the Conservancy from pursuing its mission are vanishingly rare—as well, it goes without saying, as unsuccessful.

There is surely a need for conservation organizations that will confront environmental miscreants on every front, for organizations that are not pragmatic in the pursuit of their goals but instead are idealistic both in mission and method, for organizations that will insist that enforcement officials uphold and polluters observe environmental laws, for organizations that will use the courts, if need be, to compel them to do so when politics, inclination, budget, or economics make them disinclined. The environment needs such organizations. There isn't a need for the Conservancy to be one of them. It has evolved for a different niche.

On the other hand, neither the Conservancy nor any other environmental organization has earned the right to be comfortable. The ark was a good metaphor for an organization of limited resources and sensible aspiration. But documented in the same source as the ark is the admonition that "unto whomsoever much is given, of him much is required." The Conservancy's resources are far greater than they were in 1981, when *Building an Ark* was published, and it must, on behalf of itself, its supporters, and society, aspire to more than it did in 1981. While we were building an ark, the floodwaters were rising. Now we need to think beyond the ark. We need to find ways to direct those waters into manageable channels and to keep not just a few carefully chosen areas, but big parts of our land, and the biodiversity for which that land is home, from being inundated.

CHAPTER 4

Earth, Air, Water, and Fire: A Ten-Minute Case for Conservation

The planning methodology described in part II of this book can be used to guide all kinds of natural resource and land use management decision-making processes. It was developed, however, for use in conservation, specifically biodiversity conservation. The examples given are drawn from that experience. To many readers, that will be a familiar and meaningful context. For the sake of those whose experience is in different realms, to support the assertion of the previous chapter that the floodwaters are rising, and to make clear the particular conservation perspective from which I am writing, it seems important to review the case for conservation, with an emphasis on biodiversity conservation.

Humans have long tried to understand nature better by imposing classification schemes on it. Kingdom, phylum, class, order, family, genus, species has proved useful, but it lacks the universality and simplicity of an older system developed by the Greek philosopher Empedocles. He and educated people for a thousand years after him considered the basic elements from which all matter is composed to be earth, water, air, and fire. The effects of two forces, "love" and "strife," on varying combinations of the basic elements provided, for Empedocles, the rest of the answer to the question of why things are the way they are. We could probably gain a lot in our consideration of nature today by analyzing love and strife. There is, therefore, an appealing poetry in borrowing from the ancient Greek system a structure for examining the state of nature and the need for conservation in the late twentieth century. We'll consider the elements in reverse order.

If we on earth have an inexhaustible supply of anything, it must be fire. The most fundamental sources of fire, the sun and the earth's core, don't seem to have been affected by humankind's activities. For the present we don't need to worry about conserving the fire in the sun. There are periodic conservation issues with respect to the use of near-surface geothermal resources, but the fire at the core of the planet seems to be beyond our reach. What fire we can get our hands on, though, we're using up like, well, like there's no tomorrow.

The energy of the sun and that of the earth is stored in a variety of carbon-based substances. We have been burning the most accessible of those, wood, for a goodly part of the time we have been on earth with consequences that have sometimes been locally disastrous, but we're not in danger of eliminating the resource on a global scale. That's in part because as our appetite for fire energy began to get really big, we found sources that were a little harder to get to but burned hotter.

Perhaps the most wonderful of them is petroleum. Of that compound we have, with technologies that are primitive by tomorrow's standards, created everything from medicine to plastics. Yet we use more of it for fire than anything else. We burn every day in our factories, our furnaces, our vehicles, and our vessels the production of eons. The differences of opinion that exist with regard to remaining amounts are trivial. We'll run short of petroleum in the next century. What is left will be increasingly expensive, and the opportunities our children would have had to make wonderful things of that compound will be deeply compromised.

The supply of natural gas is comparable to that of petroleum. There's a longer supply of coal, but there are environmental problems associated with its recovery as well as its use. There is no need to talk seriously about peat or wood supplying a satisfactory substitute for the hotter fuels. There's a clear case to be made for worrying about the conservation of the fire on the planet.

The effects of the burning we're doing bring us to the air. Nature itself generated the complex, multilayered envelope around the earth that shields us from the worst of the fire power of the sun and, in its lower layer, restores the life to our blood with every breath. We are these days in the midst of performing several modest experiments with the air. Over a century and a half, we have converted enough fuel into fire to raise by a third the relative proportion of carbon dioxide in our atmosphere. There's some debate about exactly what the effect of a continued increase would be, but there can be no serious debate that it will have an effect that is, at best, likely to be disruptive. It will affect the climate, probably causing an overall warming and changes in precipitation that are unpredictable at a local scale. Because sea level rise is thought to be one of most likely effects, island nations are particularly concerned. No one has reason to feel comfortable about maintaining our current pattern of pouring waste carbon dioxide and other greenhouse gases into the air.

Indeed, our experiment has some design flaws having to do with the number of variables we have introduced. In addition to the gases we're adding to the atmospheric mix, we're changing the mix of plant life on earth that uses them, and we've sent some chlorine into the upper atmosphere that has a voracious appetite for the oxygen that is up there.

Some of that oxygen is in a form that is not too common in the part of the atmosphere we breathe. In fact, around where we live, that highly reactive form, composed of three oxygen atoms and called ozone, is a pollutant that we've managed to increase, to our respiratory misfortune. But in the upper atmosphere, where it has made life possible by filtering out just enough of the sun's radiation, O_3 is ready prey for the chlorine that is released when the sun's unfiltered light hits the chlorofluorocarbons we have used for refrigerants and propellants and in the making of insulators. The reaction that occurs converts ozone to chlorine oxide and O_2, the latter of which would be fine for breathing if it were in the lower atmosphere but isn't so good for dealing with ultraviolet solar radiation of the B type in the upper atmosphere. Back on the ground, most living cells don't want more solar B.

After a good long run at destroying ozone, chlorine in the upper atmosphere will find a methane molecule, and hydrochloric acid will be the product of the meeting. When that compound sinks into the lower atmosphere it will dissolve in water and come back to earth as acid rain—with the nitrous and sulfurous acid rain that result from other stuff we send into the air. To cite these problems is not to imply that they are being ignored; international action—the Montreal Protocol—on chlorofluorocarbons proceeded, for example, at an impressive pace given the scope of the challenge. Neither, though, have our problems with conservation of the air been resolved.

The return of chlorine to the lower atmosphere as acid rain brings us conveniently from air to water. Rain of a little higher pH than that to which aquatic life has become accustomed has made reflecting pools of some lakes people once liked to fish. It has also denuded some mountaintops and diminished forest vitality over a wide range. Of equal concern is what we're doing with the water resources that have been stored underground over the millennia. In parts of the Great Plains and in a number of other places around the world, mining is a good name for it. Maybe a relatively few years of intensive agricultural production where there isn't the surface water to support it is worth using up the resource that makes it possible. It'd be a good thing, though, if we didn't sterilize the soil with the buildup of salts leached out of the water we're running through it. Vice President Al Gore has said that more than 30 percent of the world's potentially arable land has been compromised by the buildup of excessive concentrations of salts.

We have shown ourselves to be plumbers whose reach exceeds our grasp when it comes to supplying water to the millions of people who want to live in desert locations that could reliably supply water for perhaps thousands; sometimes we have collected surface water for that use, and sometimes we have looked to underground waters. It's an expensive business with fatal consequences for a host of water-dependent organisms. That is, however, only one of a long list of environmental insults our waters have suffered. The waters of North America are sufficiently degraded in quality that the freshwater fauna of the Continent are its most endangered class of organisms. Twenty percent of North America's native fishes, 36 percent of its crayfishes, and 55 percent of its mollusks are now extinct or endangered.

The habits of commercially interesting organisms of the ocean are sufficiently well known and our equipment for catching them sufficiently good that we can sweep the world nearly clean of species we favor. Robert Jenkins (whose underrecognized vision and Natural Heritage systems for the conservation of biodiversity, described briefly in chapter 3, have perhaps had more positive impact than anyone's) tells of a time when a law in Massachusetts proscribed the practice of feeding lobster to indentured servants more than five times a week. The fabled New England fishery has collapsed; some tuna are so much in demand that a fisherman's recent pronouncement that they're too valuable to live makes sense in a perverse kind of way.

The shell fisheries that once thrived in the various Gulf and Atlantic bays are as often closed as they are open—polluted waters make the filter feeders unfit to eat. Surely enough has been said to establish the point that conservation action—supported by thoughtful planning—is needed with respect to our waters.

And earth? Recent estimates of the effects of humankind's earth-moving activities suggest that they equal those of the wind, water, and ice that heretofore dominated the shaping of the surface of the earth. We and volcanoes are the only earthly forces that move stone uphill. It is thought that we move more than 100 million tons of dirt and stone every day through our actions, intentional and unintentional. We dug or regraded foundations for 650,000 houses in the United States alone in 1992.

And of the magic earth, the earth that provides the base and nutrients for the corn, soybeans, wheat, oats, and rice that feed much of the world, we almost certainly lose as much by weight through erosion each year as we generate in grain. In relatively flat Indiana, the figure is three times as much; Wendell Berry, whose thought and writing reflect wisdom about humankind's relationship with nature and land, says it's twice as much in Iowa and worse elsewhere. Again, calling our use of topsoil "mining" is less a metaphor than a way to make clear the known implications of practices that have become so familiar that we don't see them for what they are. The topsoil with which we have been endowed by the processes of hundreds and thousands of years is being expended in the activity of decades. The cost to the quality of our waterways and bays is salt (or, literally, silt) in an open wound.

If it isn't erosion, it's desertification by tree removal or sterilization when the often thin topsoils of tropical forest are exposed to direct pelting by tropical rains in the absence of the luxuriant vegetation that so long sheltered them. We have a lot of ways to destroy the earth directly.

We're doing as badly with the plants and animals that live on or in the earth, and the consequences are perhaps even more troubling. The estimates vary, but whether it is ten per day or ten per hour that are being snuffed out, we are clearly in the middle of one of the most devastating plagues of extinction that has ever befallen the earth. Scientists cite the end of the Cretaceous period, about 65 million years ago, as the most recent comparable cataclysm. The instant crisis is, of course, not comparable; indeed, it is fundamentally distinguishable from all others by one critical attribute. It is the first such event that has occurred because of the choices being made by organisms living through the middle of it.

Because this extinction event, as a symptom of the destruction of the natural world, is especially compelling to me, because it can serve as a microcosm for the case for conservation in general, and because biodiversity conservation as a mission was the driver for the examples provided in this book, I will add a few lines to the sketches that have characterized the conservation outline provided thus far.

Extinction, it is sometimes pointed out, is natural. Indeed. The fossil record reveals a couple of handfuls of major extinction events that have determined in part the complement of organisms with which the earth is now supplied. Recovery of species diversity after major extinctions, geologists say, occurs rapidly—it takes 5 to 10 million years. This is a short span of time for a system that measures its age in billions of years but one of little meaning for the relationship of humanity with nature. Humankind, which appears in the fossil record of the Earth in its earliest recognizable form only a couple of million years ago, has done nothing to discernibly increase the rate of speciation, though we have shown some facility at propogating within species that are beautiful and useful to us, the traits we like the most. We have, on the other hand, by our activities increased the rate of extinction to something between a thousand and ten thousand times the background rate.

It has been argued that humans are natural, that we were equipped to become and have become the dominant species on earth (indeed, Harvard biologist E. O. Wilson says that no other single species of large land animal has ever existed in such abundance) and that this extinction spasm must therefore be natural. Nothing to worry about, unnatural to try to avoid.

It is our capacity to be concerned (or not) that makes it inappropriate to contrast one future with another by use of the word "natural." It is within the nature of our species to treat the environment as an opponent to be tamed, conquered, or subdued. It is also within our nature to treat the environment as the source of our sustenance, comfort, and prosperity; our home, but greater than our home. A thing to be cherished, blessed, cared for, wondered at. We can choose which of our natures we will allow to guide us.

Our relationship with nature is complex. The Japanese are, as a people, often said to be lovers of nature. What shall we make of such an assertion? We Americans might say that about ourselves. Watch what we do, though, not what we say, what we paint on our cave walls, what we put within the bounds of our gardens. As long as humankind has been able to make choices—probably since before the dawn of agriculture, some 8,000 years ago—we have chosen the destructive path. The habit is strong in us. But treading now along the way we took when nature seemed endless and when there were fewer of us will result in the elimination during the next century of perhaps a quarter of the organisms with which we share the earth. A word about the number of species on earth, before addressing the consequences of the loss.

Scientists have named and classified some 1.5 to 2 million forms of life on earth. No informed person doubts that many more exist; the estimates run from 5 million to 100 million. It's certain in any event that we are losing species daily that we have never even managed to identify. If it is undesirable to lose such species, the need for action to avoid the losses is urgent.

We can take the measure of our relationship with other organisms in several ways. Each of them has meaning for us but even more depth if considered in the light of generations to follow us. Surely a part of the meaning of our existence is bound up in the existence of our literal and figurative children, though we act in so many ways as if we owe them nothing. Surely we have a duty to leave them a world with as many choices, as much variety, as we can—or perhaps even as we had. We are instead spending the inheritance we received with little regard for the impoverished legacy that will be the result. Every day makes a difference.

If we choose, we can leave a world in which those who follow us can still feel the spiritual refreshment that comes from experiencing the wondrous and diverse plants and animals that share the world with us, and the marvelously wrought manners in which they are combined. We can leave a greater, rather than smaller, number of organisms that may be useful to humankind. It is estimated that forty percent of our prescription medicines, for example, utilize extracts from wild organisms. We have not screened most of the known organisms for medical utility, we have obviously screened none of the greater mass of living organisms unknown to science, and we have screened no organisms the way tomorrow's scientists will be able to screen them.

We can, if we choose to so shape our behavior, preserve wild and therefore frequently hardy forebears of the plants we've domesticated for food and other purposes so that resistance to the plagues that afflict inbred mono-cultures can be bred back in. We can save the wild and potentially renewable sources of plastics and oils and industrial chemicals.

We can leave in the world the redundancies and supports that make natural communities strong and resilient. We can leave in place the pest con-trol mechanisms that cost nothing; recent research, for example, suggests that birds perform that service for forests.

The prima facie case is made: we ought to conserve the animals and plants that share the globe with us. Were it necessary to make the argument stronger yet, there are, still uncited, the ecological services that assemblages of species perform in combination—soil formation, water purification, cli-mate control, to begin. And there is another line of argument to be followed that begins not with humankind's relationship with nature but with the intrinsic rights of nature and humanity's obligations thereto. The prima facie case is made, but there is a response. Saving species requires that humankind govern its actions with that objective among those it keeps in mind—that we forgo opportunities that provide us other benefits in the short term. The choice to conserve comes with costs as well as benefits.

It would be convenient if we could avoid some of the costs by pre-serving species in zoos and botanical gardens or through the application of technologies like cryogenics. Aside from the philosophical question of whether a tiger is really a tiger when it is in a zoo, there are two other diffi-culties with that strategy. We need zoos and botanical gardens. They perform a useful role in education and in recent years have substantially expanded their conservation activities. But maintaining species outside their natural contexts is tough, expensive business that will not tolerate even momentary lapses in human attention. Also, it's almost unthinkable that we could sup-port in deep freezes or captivity as many species as would need such support if that were the principal conservation strategy, even if doing so were a simple task. Maintaining organisms in the wild is far more certain, likely to be less expensive, and generally is forgiving, if not appreciative, of the absence of human attention. Zoos and botanical gardens and deep-cold technology will not be substitutes for in situ conservation—conserving organisms in context, where they live.

As an isolated or intellectual proposition, the choice is obvious, the benefits great. We ought to save species. The proposition, though, is not iso-lated. Examine the costs. Wild organisms exist in a real and bumptious world, and they run into humans, increasingly, everywhere. To save species, we have to ask ourselves whether this housing development shall be stopped, whether this city shall look elsewhere for water because this particular species or two (or many, if it's a particularly rich environment) may be threatened or eliminated—this particular one that we know of no use for, that there are others much like, that only a biologist could love. We have cave invertebrates threatened by drawdown of the Edwards Aquifer, a critical element of the water supply for Austin and San Antonio. We have gnatcatchers in the path of development in southern California. We have one of several species of western squawfish troubling dam builders in Colorado; we had big science stymied by the presence of a subspecies of red squirrel. In this real world, an understanding of the critical importance of species conservation as a general principle has to be pretty firmly in place or the particular species loses the argument every time.

Alice Rivlin, director, at the time of this writing, of the Clinton administration's Office of Management and Budget and once a member of The Wilderness Society's board of directors, told a group of Nature Conservancy staff members a few years ago that we reminded her of advocates for health research. To them, no commitment of dollars for research seemed to be enough, research was needed on a seemingly endless list of subjects, and it was difficult or impossible to get them to choose among all those needs. To us, every species had to be saved. Not a reasonable position, not a *useful* position, to a politician. To her there had to be a better way to answer those who would argue that a development was more important than the species that stood in the way.

I wasn't willing at the time to concede that the imperatives of saving them all weren't strong enough. The general principle was firmly in hand. Accepting less, was giving in to a strategy of divide and conquer; we'd eventually lose every argument. Then too, who but a god knows which plant or animal we need or will need and which we can let go? Some religious traditions might suggest they were all put here for a purpose.

My reaction was right—and wrong. We really ought to aim for the conservation of every species, and every concession wrongly suggests, implicitly, both that we don't need them all and that we can somehow say which just aren't worth the cost of the choices we have to make to save them. But Rivlin wasn't challenging that reasoning. She was reminding us that politics was going to play a big role in resolving conservation questions and that if we wanted to help shape that role, we had better attend to the matter of developing advice that would be useful to a politician. The case made earlier for a conservation response to the biodiversity crisis acknowledges by its claim of urgency that the arguments conservationists have advanced to date aren't useful enough: there is a crisis; we aren't winning. Instead of holding steadfast in the face of Rivlin's complaint, we might choose to acknowledge the certainty of losses and marshal our resources to save the best array of species we can save.

*   *   *

A Stanford biologist, Paul Ehrlich, wrote an article with E. O. Wilson several years ago that suggested a moratorium on the development of undeveloped land in the United States. Such a resolution to the problem of preserving biodiversity in the United States doesn't appear likely, as Rivlin reminded us, but, interestingly, Ehrlich and Wilson paired with that suggestion a call for ecologically conscious development in less wealthy countries. It is clear, therefore, that they were aware of politics and the reality that goes beyond politics: there is a range of human needs and desires that has caused us to compromise biodiversity. They acknowledge the aspiration of developing countries to raise their standard of living and to do so at some inevitable cost to wild and open spaces. In the United States we are, in their view, wealthy enough that we can afford not to develop any more undeveloped land. It isn't necessary to decide whether they are right or not. We are surely wealthy enough that we don't need to destroy any of the species that remain in this country in pursuit of other kinds of wealth and security. That, therefore, should remain our stated goal here and should be part of our answer to Rivlin. If we care about species elsewhere, we in the United States erode the foundation from which we want to argue for their preservation if, with our resoundingly higher standard of living, we say that we cannot afford to save them here.

If we want to save the most species elsewhere, we'll focus on tropical forests and coral reefs (because there are so many species there) and island ecosystems (because so many of the species there are endangered). Perhaps, as our knowledge grows, we'll add other foci; some in the Pacific Northwest direct our attention, for example, to the lesser-known forms of life that thrive below the ground in their part of the world.

But is preserving the most species the right goal if we acknowledge that worldwide we won't save all? The most may not be the best array. Some conservationists have begun to look at the family trees of species—the science of cladistics—with an eye toward getting the most from what we are able to save, from what we have the resources to save. If, to use a hypothetical example, many of Darwin's finches were endangered and the species from which they emerged was still in existence, we preserve many genes and many possibilities by preserving that species. And we might, as a second strategy, take the same cladistic information and preserve the species that appears to have evolved the furthest from its ancestor, thus working both root and distant branch of the family tree.

Others suggest that we first see to providing secure conservation for representatives of each family, or genus, before we go for multiple representatives of species within these higher classifications. Kent Redford, director of science for the Conservancy's Latin America program, has pointed out that if we can save only a limited number of mammals and we don't allocate our resources preferentially among them, we will end up saving a lot of rodents and bats, but we might never get to rhinos and pangolins. Carry the logic and the question forward to the entire animal kingdom, and our portfolio will probably end up loaded heavily with beetles. That might be the array we want, but it is worthwhile to think about different kinds of approaches to the allocation of our limited resources.

Conservation on the ground doesn't and cannot proceed as such an orderly march that any prescription for priorities is likely to be precisely filled, but the general direction in which we choose to head makes a big difference. We might, for example, if we departed from the "most species" goal, spend relatively more of our time in the Arctic, or on the chaco in South America, or on the steppe in Europe and Asia, or on the prairie in this country. In any event, we owe some additional thought to which losses of species hurt most, and not because we ought to countenance society's desire for an easy way to avoid hard choices. Not, either, because we ought to supply society with a salve for its conscience when it is unwilling to allocate the necessary resources or modify its activities enough to leave room for species other than humans and our yard and garden wards. We should consider Rivlin's point a challenge and respond seriously because we need to acknowledge that like it or not, society is making and will continue to make choices that result in loss of species, and we can either use our influence to direct some of those choices or watch the game, feeling righteous and secure in the intellectual superiority of our absolute position, from the sidelines.

The tools we use will be essentially the same whatever philosophy guides the choices we make with respect to priorities and, indeed, whether it is biodiversity we want to conserve or forest, range, water, or soil. Conservation that will last can be achieved only if we are able to maintain the ecological processes that sustain the resources we are working to save. There is a converse dependency as well, which will reveal itself upon the loss of a critical species or community or a critical number of species or acres of community or volume of the resource. Maintaining ecological processes requires an ecosystem approach—usually, but not particularly usefully, called ecosystem management. That, conveniently, is the subject of the next chapter.

CHAPTER 5

Ecosystem Management

Conservationists in recent years have rallied around an approach to the work that goes by the name of ecosystem management. It has become the great new hope, the enlightened path. All things considered, that's good. It isn't that people who use the term mean anything consistent or reliably definable by it. Rather, like the near-universal bow now made to sustainable development, some good attaches to the use of the words—some vestige of the meaning they had before they took on a life of their own—that has the effect of moving the user toward making better, more thoughtful, more conscious decisions with respect to the uses of lands and waters.

If an age ever arrives in which humans really can manage ecosystems, we will have come a long way. Plant ecologist Frank Egler said, paraphrasing British geneticist J. B. S. Haldane, that ecosystems are not only more complicated than we think—they are more complicated than we *can* think. Despite all the talk, in other words, we aren't managing ecosystems now. In rare instances, we are taking the first unsteady steps toward that worthy goal by thinking about the ecosystems that exist in the places we live, work, exploit, or try to conserve. It doesn't take long before we prove Egler right, but the attempt itself is worthwhile and leaves us with a richer, if inevitably incomplete, understanding.

An ecosystem is a real thing, but not a very classifiable thing. Most knowledgeable people who use the term would still say that it refers to an ecological community and its physical environment. Simple enough. But the cases that fit that definition range from the "rivers, coral reefs, and ponds" of Otto and Towle's *Modern Biology* to "a jar with a fish, snail, water plants, water, and sand." An ecosystem cannot, given the textbook examples, be what common usage would suggest—a conservation objective that fits neatly near the top of a biodiversity hierarchy that begins with gene and proceeds through population, species, and natural community before arriving at ecosystem. It's convenient shorthand to say "ecosystem" when we mean a large, somehow coherent piece of the land and waterscape. But that convenience has a cost, and the cost is that none of us can know what another really means when work is described as ecosystem management.

The Environmental Protection Agency, for example, has recently been trying to achieve some synergy in the administration of its programs by organizing around what it has chosen to call ecosystems. The EPA is a big organization with a big mandate. When its management thinks of ecosystems, it thinks big: the Gulf of Mexico, the Great Lakes, the Great Plains. There is surely good to be done by making discretionary decisions about where to concentrate programs in a coordinated way. In their enthusiasm for this sensible idea, some have called it ecosystem conservation. It takes more than a common geographic agenda to make that label mean something. The programs that would be needed to make a fundamental difference in ecosystems of the size the EPA has chosen would have to be of the same scale as the forces that shape and affect those systems: continental scale. Perfect enforcement of the laws the EPA administers and perfect grant making under the programs the EPA manages will not conserve those systems. It is more than a trivial task merely to state what is meant by a conserved system when the system is the Gulf of Mexico. Our reach and our grasp are more than adequately stretched by the challenge of conserving the Mobile Bay ecosystem, a minor notch in the sweep of the Gulf shoreline, which is the outlet for the drainage of a fair piece of land all by itself—the watersheds of the Alabama, Tombigbee, Mobile, and Tensaw Rivers.

When others in government talk about ecosystem management, they seem to be using the words to describe the need for various federal agencies that share common boundaries to coordinate the management activities they undertake within a defined land area. It isn't easy, however, to productively coordinate management activities when the management plans to be coordinated are founded on markedly different premises. The first question of coordination becomes "Toward what end?" If the answer is any other than "Toward the greatest natural integrity that can be achieved," the management plan that emerges isn't really designed to conserve an ecosystem, and we've arrived at what is usually happening under the label of ecosystem management.

Most often, what has come to be called ecosystem management is simply an effort to apply something of what is known about the way an ecosystem operates to a quest to obtain some desired good or service that the ecosystem supports, creates, or maintains. A better phrase for this than "ecosystem management" is "taking an *ecosystem approach*" to management for a product of the ecosystem that is known to have some value to us. The clarity of purpose at the heart of forest, game, or grassland management methodology as practiced through most of this century is appealing, but it has, as applied, put us in a frame of mind to act as if the object we seek exists in isolation. Nothing in the natural world does, and we ultimately pay when we ignore that rule. A forester trying "ecosystem management" thinks not only about trees but also about water, soil, nutrients, pollination, insect damage, and fire. The objective, though, is usually to better and more securely produce trees. And for foresters, as well as for conservationists, city planners, cattle growers, or developers, an ecosystem approach will almost certainly produce a result that can be sustained longer and will produce fewer negative effects than the single-minded, good- or service-focused approaches many of us have traditionally used.

If ecosystem management, therefore, usually means—something good—thinking more broadly about the natural support and maintenance of the resource a person or entity means to steward, what it does *not* usually mean is governing activities and planning conservation actions purely to restore or maintain integrity in a sizable and somehow coherent piece of the landscape. Ecosystem management does not, in other words, mean ecosystem conservation. Given Egler's point and the breadth of the definition of the term, that isn't surprising; we don't very well know how to conserve an ecosystem anyway, and our inclination and sometimes our needs drive us to simplify and specify discrete rather than encompassing objectives we desire to secure from the area we intend to manage.

That being the case, the vision being promoted by those eager for an answer to environmental conflict—of ecosystem management as a new and relatively simple solution to all of our conservation problems—is a false one. But it is an alluring one. It has, among some enthusiasts, for example, made single-species conservation unfashionable. The notion seems to be that if we manage ecosystems, we needn't worry about the species within them. If ecosystem management meant holistic ecosystem conservation, that idea would probably be more correct than not, and it would be clear that as a resource allocation preference, we usually ought to set our sights above the species level. But because most use of the term doesn't mean that, I believe we haven't just yet reached the point at which we can strike species-by-species conservation from our conservation agenda. That isn't to gainsay the argument that the gamekeeper's approach to conservation has occasioned notable harm to ecological systems. But a single-species conservation plan need not ignore biological context. We can instead adapt for a single species the usual meaning of ecosystem management and take an ecosystem approach to conservation of a rare species. Such a methodology approaches the rare species holistically, explicitly acknowledging that it is either dependent on or, at least, most secure existing within, the matrix of biological and physical entities that naturally surround it. To conserve the species, in other words, one must address not only the species but also the ecological processes that set the stage for it and support its existence.

The Nature Conservancy has in recent years been evolving toward taking an ecosystem approach to biodiversity conservation. That evolution saw its first explicit expression in a strategic plan formulated in 1990. The plan called for special attention to the protection of "whole ecological systems." The Conservancy was so deeply attached to the establishment of nature preserves as a biodiversity conservation tool that the ecosystem approach was at first widely perceived within the organization as a distinct and separate new initiative—though the Conservancy's own preserve design manual had presaged the idea a decade before.

That link and others to disparate previous discussions were soon discovered, and the organization began to move toward integrating an ecosystem approach into all its conservation work. Some of the Conservancy's preserve managers, in particular, see the ecosystem approach as a way to answer some troubling questions they had long been asking, most notably how to maintain rare elements of biodiversity when the ecological processes that naturally sustain them encompass more land than is included in the preserve.

But there are even more powerful reasons to move toward an ecosystem approach to conservation than achieving better conservation results at individual preserves, and those reasons emerge from a strategic analysis of possible tools and methods for biodiversity conservation. Relatively few elements of biodiversity can survive in the long term outside the system context in which they evolved, absent constant management attention from humans (I'd bet on cockroaches, rats, coyotes, and dandelions). If nature preserves are not designed to encompass the ecological processes that sustain the plants, animals, and natural communities the preserves are supposed to protect, identifying and providing the attention needed is not only almost as challenging as maintaining those elements in zoos and botanic gardens, it is also nearly as expensive and never ending.

Because that is so, establishing nature preserves of the relatively small size that characterizes most preserves in the United States, while good for our hearts and souls, cannot and will not by itself result in the preservation of as much biodiversity as we (certainly conservationists, but also society in general) want and need. The monumental effort of supplying, with our resources, for even a relatively few preserves the ecological context a larger landscape would have supplied for free will exhaust us long before we reach our biodiversity goals.

Establishing preserves of any size is nevertheless an attractive conservation approach: the conservation result obtained is tangible, and the legal action taken is lasting. That's pretty good, but not good enough. We need the conservation result, not simply the legal action, to be lasting. Achieving that lasting conservation result outside the ecosystem context takes enormous resources at each discrete preserve, and pursuing biodiversity through preserve acquisition requires the establishment of a large number of preserves: all in all, a recipe for a failed strategy. The challenge, then, is to make conserving the ecological context—an ecosystem approach to conservation—not simply an additional strategy, and not just an alternative principal strategy, but a more successful principal strategy than establishing nature preserves.

It is evident, first, that we cannot effectively pursue an ecosystem approach to conservation if we do not use tools that extend beyond the acquisition of land. The lands and waters encompassing ecological systems tend to be large, too large for purchase to be the strategic solution very often. When the Conservancy realized that, it was surely perceived as a challenge, but it was also a gift. It required the organization to begin to think about conservation strategies that involved activity beyond land acquisition. Conservation possibilities that had always existed but that had seemed unapproachable presented themselves: aquatic organisms, migratory birds, the Everglades–Florida Bay–Florida Keys system, and many others.

Acquisition of key areas remains an important tactic for biodiversity conservation and for the Conservancy. But working with owners of lands not acquired has assumed new importance; in fact, it has often become the central objective of a different strategic focus. When that focus is conjoined with a commitment to export the lessons learned, protecting whole ecological

Figure 5-1. Freshwater mussels near Pendleton Island, Clinch River, Virginia. © Steve Croy

systems can become the means to a new strategic end as well as an end in itself. That new strategic end is protecting ecological systems without intensive investment by and involvement of professional conservation organizations—conservation by trading stories that spark new action. And if the fate of biodiversity depends not on the system of nature preserves we establish but, ultimately, on our success in establishing preserves *and* using a variety of other tools to save a critical amount of natural quality in the landscape as a whole, that new strategic end is of critical importance.

The leverage implied in the new strategic end is also of critical importance if our interest in conservation is not so much biodiversity but the life support systems for humans that nature supplies. Peter Raven (director of the Missouri Botanical Garden) among others, has suggested that it is the maintenance of those services that will attract the level of support we ultimately need for conservation. That may be right; people love nature, but society values most fundamentally ecosystem services—soil formation; conservation, cleansing, and regulation of the flow of water; oxygen contribution and carbon dioxide absorption; climate control—the obvious contributions of nature to our survival and our economic well-being. Ecosystem management connotes the preservation of those fundamental services, but, again, probably cannot deliver on that promise unless maintaining those services becomes the objective of an ecosystem approach, and it will not deliver on it anyway unless a very great amount of land is influenced, directly or indirectly, by ecosystem conservation efforts.

The best current ideas for generating that kind of influence involve conscientious efforts to document and disseminate land use lessons learned—new and old ideas for living in harmony with nature. People making land use decisions need ideas for satisfying their economic and environmental aspirations at the same time and on the same land. The knowledge we accumulate in pursuing an ecosystem approach to conservation can be adapted to facilitate the efforts of others concerned with conservation and the future of their communities—the planners, the corporate leaders, the community groups, and the farmers who will, collectively, determine the fate of the larger landscape.

It is reasonable to hope that through dissemination of lessons learned we can gain leverage from our intensive conservation efforts, and there is a strategic need to do so. Even so, success is not certain. Whenever possible, therefore, we ought to establish the field projects in which lessons will be learned at the most biologically important places on the earth—we ought to preferentially select the intact transect of southwestern natural communities of the Gray Ranch, the upstream refugia of big-river mussels of the Mississippi River system, the migratory bird congregation sites on Delaware Bay, the coral reefs of Palau, the eastern slope of the Andes. If we are not successful in gaining leverage toward conservation of other sites from our work at such critically important sites, we will at least have stemmed the tide of loss of the natural world in some of the places where it matters most.

*   *   *

The lessons learned at these places will have relevance to the broader landscape because successfully implementing an ecosystem approach to conservation in any location is nearly certain to require more than attending to biological and physical phenomena. Conservation as we near the third millennium will have to be supported by a tripod. Perhaps the main load-bearing leg is ecology. But stability will be achieved only with the addition of two other legs, community and economy.

The dictates of ecology supply the organizing and limiting principles governing conservation action, including action with human communities and economies. To do good conservation, practitioners have to develop, maintain, and improve their understanding of the relationships among organisms and between organisms and their environment. In almost every case, understanding of the ecological processes relevant to any natural system will be achieved incrementally, and the effort to gain that understanding will be ongoing. For a conservationist, maintaining the integrity and viability of ecological processes is a foundational objective.

There are ecological systems that can be sustained through conservation of a relatively small piece of land—a midwestern shale barren or limestone glade, for example, which (I say at my peril) may be adequately conserved in an area of fewer than a hundred acres. Such systems represent the rare instance in which attending to the other legs of the tripod is unlikely to be critical in the short term, at least in a society that is comparable in wealth to the United States. But relatively few natural systems—probably none, for example, that depend on moving water—can be well conserved at such small scales. And when taking an ecosystem approach to conservation requires attention to areas of land much bigger than a hundred or a few hundred acres, the short-term effect of conservation on the economy and the converse become issues. As well, the support of the human community for conservation, always a desirable goal, is likely to become essential. Real, lasting conservation nearly always requires community understanding of, pride in, and responsibility for the natural systems that surround and, after all, support the community's social and economic systems.

A destitute human community cannot and will not fulfill those responsibilities to the natural system in which it is embedded. And economic growth and opportunity remain important issues even in communities that are quite prosperous. At the least, therefore, effective conservationists must, as will become clear in chapters 8 and 10, understand the economic drivers of the relevant human communities and the principles that characterize vibrant local economies. Where there are others working in the community who have economic development as their principal objective, the conservation imperative is to make it possible for them to meet their goals in a manner that is consistent with conservation goals. Where there are no such possible collaborators, economic issues must still be addressed; economic problems must still be solved. They join ecological and community issues on the conservationist's ecosystem approach project list.

Workable solutions will accommodate people's need and desire to make a living on, in, and around lands and waters of conservation importance and will do so in ways that are consistent with or even reinforcing of a sound and healthy environment. When the work people are doing destroys natural systems, that ought to be seen as a challenge that calls most urgently for a search for new economic ways and means. People need work, but an environment stripped of its naturally functioning systems cannot, in the end, be a sound and healthy environment.

The key phrase in the previous sentence, of course, is "in the end." If we don't care whether our presence can be long sustained or can produce less incidental harm, a sound and healthy environment may not be worth the bother to us. And the evidence would suggest that it will take some effort to bring out the capacity in any community for caring. Most of what humankind has wrought since the industrial revolution looks as if it could have been done under the banner of John Maynard Keynes's words: "In the long run, we are all dead."

I think it is possible, incidentally, that Keynes would recoil at the frequency with which he is now cited as authority for that view of planning. Children and grandchildren are a part of "we." They are now feeling and will continue to feel the effects of our environmental decisions to a degree that Lord Keynes could have then perceived only dimly (though he is, interestingly, said to have seen overpopulation as a coming issue for humankind). "Darkest Africa" and the Wild West, at least as represented by Buffalo Bill Cody's show, were part of the world of his experience. There were still evident terrestrial frontiers. We can no longer count on finding uncut trees over the next hill, virgin grasslands upriver, abundant fisheries a little farther from shore. We can't plumb deeper aquifers. Planning for the short run cannot, given the resource situation we face, really be called planning at all.

Part II of this book presents a discipline for planning that is designed to provide direction for short-term action taken with a long-term perspective. The discipline offers a way to begin to think about the organic unit that must be addressed in an ecosystem approach to conservation. Because that unit comprises the linked systems of the ecology, the community, and the economy in a place, it is even more challenging to contemplate than Egler's ecosystem.

PART II

A Planning Discipline for an Ecosystem Approach to Conservation

CHAPTER 6

Getting Started; the Five S's

For many of us, choosing the conservation task we're going to work on isn't much of an issue. We are attached to a place by emotion or residence or assignment, and our work is to conserve that place or some resource in that place. For others, deciding where to apply available conservation resources is the first choice to be made. There is one obvious and one less obvious way to make such a choice.

The obvious way of choosing is to seek the project that holds the best prospects for fulfilling the decision maker's particular conservation purposes. If the purpose is biodiversity conservation, we might, for example, seek a large area of land in which relatively intact natural processes help secure a number of rare species and natural communities. If providing outdoor recreation is the objective, the project site would be accessible and scenic, with diverse recreational possibilities including hiking, camping, water sports, and picnicking.

A less obvious but perhaps even more important attribute to seek in a project is the potential for leverage. In comparing one project with another by this standard, we would try to decide where we could learn and teach the most about how to solve problems that represent common obstacles to the achievement of conservation purposes. We would consider a possible project's visibility or potential for it, the typicality and tractability of threats, the likelihood of political and financial support, and the presence of desirable partners.

When both the first and the second set of criteria are met, there is a priority project to be done.

Identifying projects that meet the first set of criteria calls for one set of skills; finding projects that meet the second, a different set. That is in fact the case with every step in planning for and carrying out an ecosystem approach to conservation. In The Nature Conservancy, both project selection and major projects are, therefore, often undertaken by ad hoc teams composed of members with diverse skills. Typically, team membership would include at its core a program director with broad perspective and responsibilities (Conservancy program directors have a wide variety of backgrounds, including science, business, and law) and a biologist or an ecologist. Around that core, with greater or lesser responsibilities depending on the project, would be added government or community relations specialists, lawyers, fund-raisers, business development consultants, communications experts, and natural resource or agricultural economics experts. Team members frequently reach out within their specialties for additional expert advice on discrete issues as well helping to shape a vision for the project as a whole.

If there is a choice to be made about whether or not to proceed with a particular project, the project team ought to be assembled to spend perhaps a day or a day and a half on a preliminary run through the planning discipline that is presented in summary form on the following pages. The objective of this exercise is to make a threshold decision about whether the project presents a conservation priority that ought to be pursued or, in the alternative, whether it is unlikely that the hoped-for results can be achieved, even though the project appeared at first to have exemplary attributes. The expenditure of resources on a serious threshold analysis is well justified because making the decision to initiate a conservation project with an ecosystem approach usually means making commitments that cannot be casually abandoned, and expending a large amount of resources over a long period of time. Even when a conservation team is not in a position to make a choice about what place and what conservation object to pursue, this kind of threshold planning exercise will improve the team's perspective when employed as an introductory overview to the project.

From the comfort of a position as an observer, the virtues of planning seem self-evident. To a conservation doer, facing urgent and diverse daily challenges, planning isn't always so clearly a high priority. The Nature Conservancy manages that gap in part by packaging planning in a particularly accessible way. Marketing of that kind is a familiar theme in The Nature Conservancy. For the better part of a decade, for example, the Conservancy taught its staff, its volunteers, and anyone else who would listen that it preserved biodiversity with a methodology that could be described in three words: identification, protection, and stewardship. It was a useful formulation, it was repeated constantly, and in it was a fair summary of the work TNC actually did. The simple message was good external marketing and a self-reinforcing internal management tool.

These words still describe the Conservancy's work tolerably well. Conservancy people don't use them so often now, because the words began, after so much insular use, to take on the status of terms of art. And artistic standards begin to diverge widely when used over half the globe by more than a thousand people during the course of a decade. Perhaps even more important, some of the reinforcing use of that formulation (Conservancy planning formats, for example, were organized to have identification, protection, and stewardship sections) was beginning to compartmentalize thinking as well as expression. In Conservancy practice, protection came to mean land acquisition. If you plan to "protect" a site by acquiring it, nominally protect it by acquiring it, and report that a site is protected when it is acquired, it isn't unnatural for you to think of it as protected. And, of course, as the just-concluded discussion of ecosystem management makes clear, it is probably not protected by any objective standard. The point is that the marketing benefits of the simple formulation "identify, protect, acquire" worked, almost too well.

Even so, the benefits of presenting an ecosystem approach to conservation planning as five S's—what Conservancy board member Dan Simberloff has called an alliterative mantra—seem to outweigh the costs. "Five S's" is marketing—just like identification, protection, and stewardship—and it has the flaws that marketing has. It's a little simplistic. But it captures the essence of the right message as good marketing does, and it's memorable, as good marketing is. Systems, stresses, sources, strategies, and success: the five S's. There is more to the mantra than marketing. A conservation plan that is structured around the five S's is likely to be a better plan.

Briefly, the planning objective is to understand the ecological *system* we want to conserve, identify the ecological *stresses* that burden it, trace the stresses to their socioeconomic *sources*, develop good *strategies* to address the sources and alleviate the stresses, and, finally, determine how to define *success* and measure progress. Each of those steps is introduced at slightly greater length on the pages that follow and is more fully considered in succeeding chapters.

The word "system" emerged from the Conservancy's experience as a useful key for opening the first door to thinking well about large site conservation. It emphasizes the main point of the ecosystem approach to conservation—that the object of conservation planning does not exist in isolation. That is true whether we are trying to protect a species, a natural community, or a group of natural communities. It's also true if what we are trying to conserve or sustainably use is water, soil, trees, or grass.

Defining and describing the system to be conserved seems simple enough, but step one is a good place to be very deliberate in implementing the five-S planning discipline. Clear definition of the system is as difficult a task as there is in the planning discipline. Defining the system requires more than thinking about the animal to be protected or the stand of trees to be sustainably harvested. It demands, for example, consideration of nutritional requirements, of reproductive cycles, of predator-prey relationships, and of long-term fluctuations in population and the reasons for them.

We're also called upon to know the origins and purposes of ecological phenomena that affect the system. With that knowledge, we must ultimately decide which of them require conservation attention. Finally, system definition includes thought about which things that surround that animal or that forest are not very important to its ecological health, so that we can later begin to consider what compromises to the system we can countenance.

This last set of questions will lead to the next step in the discipline, which is to consider threats to the system's continued existence or to the maintenance of its integrity. The Nature Conservancy originally called the second step in its planning discipline "threats analysis." Project teams understandably adopted "threat" as the unit of analysis. The Conservancy concluded after a time, however, that its project teams would be better positioned to generate good strategies if they considered threats in two more narrowly defined steps. Team members are now advised to ask first what the ecological stresses to a system are—independent of the source of those stresses—before separately tracing those stresses to their sources. By expressing a stress in terms of the specific effect it has on the conservation object rather than describing it, as we usually would, in its social context, Conservancy project teams have enhanced their creative potential for finding ways to avoid the stress. If we do not consciously alter our natural mode of expression, we will, for example, call a proposed road a threat in an estuarine system. We are then immediately inclined toward the conclusion that we must stop construction of the road. Threat: road. Solution: stop road. But if we separate the threat into stress and source, the stress isn't the road. The stress is, for example, loss of tidal flow. That formulation of stress inclines us to think, instead, of ways to keep tidal waters flowing through the pathway that is the proposed location for the road. The best way to do that might be to stop the road, but if that is a losing battle, the day isn't lost. Culverts may answer.

The second half of threats analysis, identifying sources, is often the most straightforward task in the planning discipline. Natural systems have, of course, often been stressed by natural sources. Infrequently and at irregular intervals over the course of millions of years, volcanic eruptions, collisions with large asteroids, and natural climate change have imposed their cataclysmic will on biological systems. Sources with comparable power to threaten ecological systems are, however, now constantly imposing similarly profound changes all over the world. These changes have social and economic origins. They are generated by humankind, and they can be managed by humankind if we choose to manage them.

Clear identification of stress helps clearly highlight the source. In the example just given, the source of the threatened stress is clearly the proposed road, though a broad-thinking planner might identify the source as the societal desire to improve transportation in a specified area—and might therefore be led to strategies that don't even assume a road. But when the system is, for instance, a stopover point for migratory shorebirds, and a critical component of the system includes the availability of abundant food in a small bay, we cannot determine which sources are important if we don't describe the stress to the biology of that bay in terms more specific than "pollution." Is it toxic chemicals or siltation, and if siltation, is it physical burying or is it light deprivation? As obvious as the advice to ask that kind of question seems, it bears giving because of the frequency with which it appears to have been overlooked in project plans. And if what we are thinking about is as broad as "halting pollution" in the small bay, we are likely to put a good deal of energy into resolving sources that aren't of the most immediate importance to our system.

Once sources are defined, the next thing to do is work on ways to manage them: to decide on some strategies that will reduce, eliminate, or mitigate the identified stresses. Some Conservancy users of the planning discipline have argued that the fourth "S" should be not strategy but social situation. It surely is critical to understand the attitudes, aspirations, and economic circumstances of the human communities affected by the conservation system we are working on. Indeed, only by obtaining this kind of understanding can we synthesize strategies that will be effective.

What kind of strategies will be effective? Nature Conservancy strategies are usually designed not to make the source of a stress into a better citizen, person, or institution but simply to address the identified stress. I don't think you can solve discrete conservation problems any other way, with due respect to John Muir ("When we try to pick out anything by itself, we find that it is bound fast by a thousand invisible cords that cannot be broken to everything in the universe"). We need to make some big changes if we are going to leave our planet the kind of place we can feel good about passing on to our children, but we can save some important parts in the meantime by picking out and working on some things as if they aren't bound to everything else.

It has been argued that less than fundamental changes only allow us to postpone the day of reckoning we need to have with the big problems. Little steps, though, may be the fastest way to learn how and why we must take big steps. The first really big step, after all, is to consider the environment when decisions are being made. The federal government is committed to do that with respect to its actions under the National Environmental Policy Act, and many state governments are similarly bound. An ecosystem approach to conservation is often about getting private landowners and county officials to make the same commitment, albeit without the attached bureaucracy and pounds of paper. The challenge is to develop with landowners and officials creative ways to make the act of considering the environment generate a result they perceive as positive. With a few such positive outcomes to show, we can be effective advocates for the proposition that the environment ought to be considered as other more fundamental and more far-reaching decisions are made.

The last major task of the planning discipline has to do with defining success and creating ways to measure progress toward it. As we think about defining success, the planning discipline reveals itself to be circular—developing a vision of success might as easily have been called the first step in the planning process; in some projects, system definition proceeds from the definition of success. At the least, the vision of success may, from its position as the last of the S's, refine the boundaries of the definition of the system. In a planning sequence that may have been perceived as linear, the sudden perception of a curve can be troubling. The way to handle any such disquiet is not to apologize for the bend but to emphasize it. The planning process can usefully be pursued as a sequence, and a description of it has to start somewhere, but it is not linear. It is essentially circular.

It is also important that we consider in advance what will constitute progress, as well as victory, when we think about success. Setting interim goals may seem so obvious as to require no special emphasis in the discipline, but the kind of goals we choose to set is important. Tangible and lasting conservation based on an ecosystem approach is most often a long-term project. Our nature, faced with a lot of things to do and an ultimate goal that is a long way off, is to measure our activity. That is a perfectly valid thing to do. The only difficulty is that many of us never get around to measuring anything else. And the benchmark we really want to attain in each of the programs we employ toward our ultimate vision is not activity but impact. That, in complex systems, is especially difficult to gauge in short periods of time, but as well as we are able we ought to direct our principal goals toward achieving impact and design our measurements to assess progress toward that end.

Systems, stresses, sources, strategies, success—each step of the planning discipline is explored in more detail in the following chapters. Together, they amount to a useful sequence for an ecosystem approach to conservation.

CHAPTER 7

System: The Ecological Unit to Be Conserved

Naming the system that is to be conserved is a likely enough starting point for a conservation effort. The first step of the planning sequence requires more, however. It calls for a definition and description of the system to be conserved that is sufficiently clear to provide a foundation for focused conservation action.

Defining the system is deciding what system to work on and setting useful boundaries upon it. If the objective is simply to conserve the lands and waters nearest us, for example, we need to decide first how near we mean. A watershed overlay on the answer to that question will serve as a reasonable first cut at system definition; system *description* is a much more daunting challenge (as will be discussed) with such an encompassing system definition. If we value most the capacity of a place to produce certain goods and services—aquifer recharge, recreation, biodiversity, quality of life—we can define the lands and waters on which we will work by asking what we need in order to sustain the production we value.

The Nature Conservancy's analytical framework for defining the systems on which it will work assumes a need to allocate its scarce conservation resources. It assigns a high priority to species and natural communities that have become uncommon (because, among other reasons, they may well disappear if no resources are devoted to their rescue) and a lower priority to open space and elements of biodiversity that are still common. There is so much work to do on relatively rare stuff that in recent years the Conservancy has seldom worked on anything else.

System description is a twofold task. We are required first to specify the living unit we aim to conserve, and then to consider the nonliving influences that are critical to it and require conservation attention. If there is a compensating virtue in the vagueness of the term "ecosystem management," it is that it is reinforcing of that two-step formulation: when we think of ecosystems, we tend to think of organizing our work around an ecological unit and the physical phenomena on which it depends.

The best conservation plans I have reviewed approach the task of developing a sufficient description by creating what Conservancy biohydrologist Brian Richter calls conceptual models. That term is meant to emphasize that this kind of model is designed for the limited purpose of providing analysis sufficient to justify conservation investment. Among scientists, ecosystem modeling usually means gathering large amounts of data, taking hundreds or thousands of measurements, and developing formulae that represent the interactions among those things that have been measured. Those individual bits of information can be organized into a portrait, in the pointillist style, of an ecosystem. The conceptual model is a quick sketch by comparison, but it will suffice for many conservation purposes.

The most important thing that developing conceptual models does is ensure that the question "How does this system work?" is explicitly asked and thoughtfully answered—to the best of our inevitably incomplete knowledge. How is the species, natural community, forest, range, or landscape maintained? How does it regenerate or reproduce? How does it respond to climatic variation and predation? What is its life cycle? How (in the case of species) are its weak, dying, and dead removed? What (in the case of natural community and landscape) are the successional patterns? How do those questions relate to the question of maintenance? Where does the medium in which the system is embedded come from? And so on.

You create a conceptual model, in other words, by assembling in narrative, qualitative bits what is known about the interplay of biological elements of concern and the physical setting in which they are ensconced. You consider the effects on the facts you have assembled of short and long periods of time, of the cycle of the seasons, and of population dynamics of characteristic organisms. As a sense of order begins to emerge, you chart the interrelationships among the narrative bits. You stare at the chart for a while, amend and append it, stare at it some more and talk about it with your colleagues, amend it again, and you have the first draft of a conceptual model.

It was in the creation of a conceptual model that it became clear to the Conservancy, for example, that the natural meandering of rivers was critical to the establishment and maintenance of the globally significant riparian plant communities of the upper Colorado River basin. The Yampa and San Miguel Rivers of northwest and southwest central Colorado sustain outstanding examples of evergreen riparian forests along the banks of their upper reaches. As these rivers, largely unaffected by a few small-scale dams and diversions, flow westward and down, the evergreens give way to deciduous and shrub riparian communities. The understandable protective instinct of a conservationist—to try to save these communities by stabilizing them—would destroy them; they are dynamic communities, always shifting, always, in some respects, young. The elegant model in Figure 7-1 didn't flow right out of the planning team members' pens when they sat down to ask how the system worked. The model building started in relative darkness and was gradually illuminated by the wiring up, through discussion, of a succession of ecological lightbulbs.

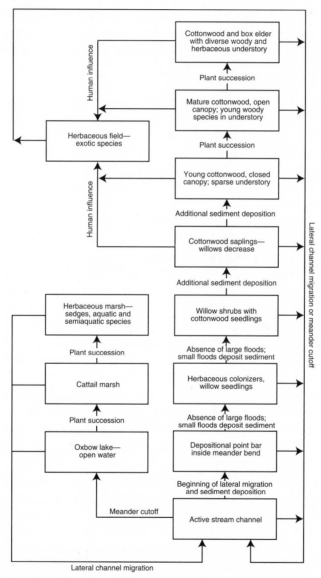

Figure 7-1. Draft Ecological Model, upper Colorado River system.

Further experience with the upper Colorado riverscapes will further inform the models; like all planning, modeling is, in its best form, an ascending series of successive approximations, informed by the experience of living and working with each iteration.

*   *   *

The use of the plural form in much of the foregoing discussion of models wasn't accidental; in most of the ecological units at which the Conservancy has chosen to deliberately apply an ecosystem approach, there are smaller ecological units within larger ones, systems within systems. For those systems that are priorities, therefore, there will often be models within and beside models. It isn't always easy to analyze clearly which systems ought to be conservation priorities, and good planners sometimes reach arguably wrong answers as they try to establish their priorities.

One of the Conservancy's project teams, for example, began with the premise that the system it would try to conserve was the lower Mississippi River alluvial plain. That was an attractive and important enough place to work, and for members of that project team it was home. They were not simply looking for an opportune place to demonstrate what gratifying results could be obtained by following a conservation discipline. They were drawn to the principal natural feature of their region.

The team used a widely understood geographic label for that natural feature as a broad system definition. It felt natural and seemed logical enough. But approaching system definitions in that way is often troublesome. Try to envision a coherent plan for conservation of the lower Mississippi and the nearby land influenced by the river—the lower Mississippi alluvial plain *system.*

The first and basic element of that system is water. The water in the lower Mississippi comes from the upper Mississippi. The upper Mississippi system is dammed, diverted, and levied. It is filled with agricultural chemicals, silt, industrial waste, storm sewer overflow, and the outflow of municipal wastewater treatment plants, some of which are working well and some of which are, well, hardly working. The upper system stretches from New York to Montana. Every natural community in and along the lower Mississippi is influenced by the quality, quantity, and timing of water delivered by the upper system. Charged with conservation of the lower Mississippi, you would probably be tempted to start by narrowing the scope of the problem with an engineering solution. You might envision some kind of incredibly sophisticated collection, treatment, and release facility just north of Memphis; one that sends downriver the nutrients, organisms, and silt load the river is supposed to carry but that holds back those it isn't. Even if such a facility could be designed and constructed, we're still dealing with a Memphis-south system that is a degraded shadow of its natural self because of the constraints, withdrawals, and effluents of ranchers, farmers, industrialists, and home builders from Mississippi to Colorado and New Mexico.

There isn't going to be a treatment facility north of Memphis. A "lower Mississippi" system conservation plan is therefore bound to deal with both parts of the system, and the project spans easily two thirds of the continental United States. Most reviewers of the lower Mississippi plan that was presented were skeptical about the prospects for success of an effort of that scope. The project team, though, was excited about a commitment to the lower Mississippi as the project focus. They had signed on to conserve the characteristics and natural attributes of the "big river"—and they found it difficult to let go of that big idea.

In the succeeding dialogue between the project team and the reviewers, the next question asked was "What defines a big-river system?" Big-river biota? The main stem would be a massive restoration project, assuming what is unlikely—that the political will to make so much as a start could be gathered. And why save that biota on the main stem? Probably all the Mississippi's native plants and animals that aren't already extinct can be found in not-quite-so big river tributaries that would be better bets as projects.

If not big-river biota, is it the plant communities of the Mississippi floodplain that define the big-river system? If riparian plant communities of the main stem river have to be saved, the first question is how to deal with the levees that line the river for hundreds of miles and keep the flood regime from being what those floodplain communities naturally received. The second question is the same one asked with respect to in-river organisms: "Cannot good examples of those communities that are more nearly intact and more easily conserved be found?"

The reason the group of reviewers put forward these irritating questions was to stimulate answers that would shed light on the issue of system definition. The answers might have revealed that the objective really was the big-river system and that daunting as it was, the real task was continental conservation. What emerged, though, was what will often emerge in such a discussion—system goals that require something less ambitious but more likely to be accomplished than a continental-scale reorganization of human and economic activity.

It became clear during the discussion that a principal goal of the project was to preserve a wide swath of bottomland forest to the north of the Gulf of Mexico and toward the middle of the continent with constant moving and periodic standing waters nearby. Forest of that kind was believed to be important for large mammals and critical, as well, for certain breeding and migratory songbirds and waterfowl. That objective presents a challenge that, with focus and resources, might be met—and an opportunity to make an important conservation difference. Even so, Conservancy reviewers questioned the decision to allocate scarce resources to conservation of the Louisiana subspecies of the relatively common black bear. There was discussion, too, about the relative conservation priority of the species of songbirds and waterfowl that would be aided. The system objectives, though, had finally been distilled sufficiently to permit identification of the lands and waters that had to be conserved to meet the objectives. With that definition, the system was of a scope that would permit the development of a sufficiently detailed description, the discovery and documentation of stresses and sources, and the creation of strategies for addressing them.

It is, perhaps, critical to understanding system definition to understand that different conservationists having different objectives could and would define, for their conservation purposes, different systems within the same geographic area. The delivery of a certain amount and quality of Mississippi water may, for example, be critical to near-shore Gulf fisheries. Creating a plan to see that needed water is delivered is within the realm of the possible. The timing of the delivery of floodwater, as well as its volume and scouring force, is critical to the function and structure of certain riparian forest communities. It is possible to imagine that certain stretches of the river or its tributaries might be reengineered to restore more natural flooding behavior if maintenance and restoration of riparian forest were the goal. To a transportation planner, quantity and constant availability of water are themselves significant. There is a set of requirements for accommodating Mississippi barge traffic that can be clearly set forth and pursued if water for barges is the conservation object. Ideal water for barge purposes might be quite different from the water regime that works best for forests, fisheries, or waterfowl. Any or all of those systems, though, might have fit within the bounds of a system called "lower Mississippi" or "big river." Without a more narrowly defined system objective such as that which was eventually delineated in the Conservancy project example, there is no way to organize effective conservation action.

I have presented the lower Mississippi plan so as to illustrate a point when, in fact, the debate over system that emerged at the review session was in part semantic. The project team had chosen to characterize the project by its geographic location because that label was easily assimilated by partners and funders. Team members had identified several systems within the Mississippi alluvial plain as priorities—including a Louisiana black bear "system," a neotropical migrant "system," and a few high-quality bottomland forest "systems." They had produced plans for conservation in each of those systems. They ran into problems when, having become used to calling the project the "lower Mississippi alluvial plain," they allowed that label, the geographic *setting* for their systems, to assume a place as a priority system in itself. They were then bound to define it and eventually propose strategies for conserving it. And it's hard to define, and even harder to produce, credible strategies for conservation of a system of such scope. Someone who says he or she is working to conserve the lower Mississippi either means something else and is speaking in very general terms or is pursuing so many conservation leads that there is not likely to be effective follow-up on any of them.

That isn't an argument for avoiding big conservation vision altogether. Our society would be better served in the long term if decisions made regarding the Mississippi were made with the goal of removing, step by step, the things we have done to alter the natural course of the river. While it isn't likely that someone will produce a practicable conservation plan that will result in the restoration of an integrated conservation ideal on the Mississippi, the ideal does represent a coherent vision. A restored and naturally functioning big-river system is a worthy goal, a useful reference point for setting strategy, and a beacon by which to steer decision making. We will make more successful plans for conservation action, though, if we address ourselves to less encompassing systems while keeping an eye on the implications of the tactics we select for realization in the long term of visionary objectives for the larger systems. The reward we get for accepting progress in smaller units is the achievement of results in a shorter period of time.

*　*　*

Sometimes defining a project of appropriate scale is less an issue than achieving focus within a project in which the appropriate scale is clear. Difficulty with focus can also be traced to lack of clarity in system definition. During the Conservancy review of the first plan for conservation of a fine central Ohio stream called the Big Darby, the planners explained that it was the faunal diversity of this unprepossessing creek that drew their attention. And indeed, of the eighty-six species of fish and forty freshwater mollusks that can still be found in that creek, at least ten are globally rare or more critically endangered. An aquatic habitat for rare species, therefore, was the conservation object. As the planners outlined the relevant system, they decided, logically enough, to begin with the watershed. They proceeded next to examine stresses and sources and lay out detailed strategies for protection of everything natural in the watershed. A good approach if neither resources nor time is an issue.

The watershed was mostly in agricultural use, but it also contained a few patches of native forest—not globally significant but of notable quality for that part of the state. In the ideal case, not only would we conserve those wooded areas, we'd also restore the rest of the landscape to the mostly forested cover under which the fish and mussels lived for most of 10,000 years. But we have a world of endangered places to attend to and less than a world of dollars, people, and time. The Conservancy chose to focus on the Big Darby because of its fishes and mollusks. The system definition was built around them, and it appears they can be conserved in an environment far more compromised than a mostly forested watershed. What they do require, however, is relief from silt and chemical runoff. That relief could be provided by replacing most of the row crops with native trees—or it could be accomplished with some adjustments in farming practices, a cheaper, likelier, and more flexible approach. A system definition of sufficient rigor for the Big Darby would not, in other words, include the remaining forest fragments within the watershed except as imminent threats to them would affect, for example, silt loading. From that perspective, creekside forest becomes important, while most nonriparian forest patches become a secondary concern. Without that kind of clarity of definition, conservationists will (with apologies) be up a creek with respect to setting priorities. Equally troubling, a definition of system that is not focused in a disciplined way will make useful strategies—like adjusting farm practices—seem to conflict more than they really do with the conservation objectives of the effort.

By this point in the discussion, some conservationists will find themselves impatient at the myopic prescription for focus. The dream of a naturally reintegrated landscape is a beautiful and useful one, and to reiterate the point made in respect to the Mississippi, ought to be held in mind as an ideal, influencing all of our decisions. But the fate of the mussels of the Big Darby and other endangered elements of biodiversity need not and therefore cannot be made to depend on realization of that dream.

We're 6 billion on the planet now, we're going to grow in number before we shrink, and we're not going to shrink to levels consistent with the logic of evolution before our species achieved dominion. Well after the last ice age, there were significantly fewer people on the entire planet than live in metropolitan New York today. The boundaries of the decisions our society will make are going to be defined by the imperatives of accommodating all those people. We are going to facilitate their efforts to eat, find shelter, and work. We are going to deal with their production of waste. These 6 billion people are, for the most part, going to insist on living pretty much in the manner to which they have become accustomed—or, if possible, better. What conservationists must do, therefore, is establish priorities among our goals, and accomplish the most essential of them within the boundaries dictated by acceptance of, if not resignation to, the society of which we are a part. We're going to accomplish our conservation goals in part by working with people to effect relatively subtle changes in work and lifestyle that will reduce stresses on systems we care about. A lot of our natural world is threatened, and only these relatively small changes are likely to happen quickly enough to preserve its foundations. Greater changes to which we may aspire will take longer to effect. We cannot allow ourselves to forget that we want to make them, but we have to attend to conservation that will produce more immediate results if we hope to have enough of the natural world left that, when the time comes that we can make bigger changes, making them will still mean something.

Preservation of the patches of woods that exist as islands within the agricultural fields of the Big Darby watershed, locally notable though they may be, becomes then not a principal conservation objective—not, in fact, an important part of our system at all—but merely a rather nice thing to do if it can be done without losing focus on the main strategies for protecting the system of freshwater fauna.

When a Conservancy project team in California was considering actions to be taken on the Cosumnes River, the issue of what's nice to do and what's a principal priority arose in a different context. An important system objective on the Cosumnes is the valley oak riparian forest that exists in the central California floodplain of the lower Cosumnes. This forest community, once quite extensive in California, has become rare over the past century. Dams, irrigation, and channelization along most rivers have recently taken from them the flood stages that pruned and fed the forest communities that had evolved along their banks. Preserving the integrity of flooding regime of the Cosumnes River and reversing the biological consequences of the much-increased patchiness and fragmentation of the remaining forest were among the principal ecological concerns. Social concerns with environmental implications included the new residential development that was displacing agriculture as the dominant land use in parts of the project area. The problems to be solved seemed clear. So far, so good.

As part of the process of preparing to choose appropriate strategies, the project team started working on a conceptual model to enhance and better document its understanding of the system objective. It turns out that Swainson's hawk, uncommon in California though not particularly rare globally, favors the river valley. In what seemed a logical enough decision, plans for its conservation were incorporated into the preliminary set of strategies. The hawk, after all, is a threatened resident of the area of principal concern.

But as a species priority for system conservation, the hawk is way down the list, considered independently. California is home to many more globally endangered species than it appears likely we will save. Why should the presence of the globally secure hawk in a natural community that is a priority vault the hawk to the top of the list for action? It isn't hard to envision actions (manipulating hunting habitat or prey populations, for example) that would have a positive short-term effect on the population of the hawk but would be neutral—a distraction—or harmful to the community that brought us to the Cosumnes in the first place.

The common conservation sense that says we ought to be managing our conservation activity with a bias toward wholeness cries out at this point that the Swainson's hawk and every other species present in the Cosumnes Valley has an ecological role in the community. The surest strategy of all for preserving a natural community would be to ensure that every part of it that we are able to identify is secure and in historical balance. But given the need to set priorities for the allocation of limited resources, we have to paint with a broader brush if the thing that brought us to the site is a natural community. If the community would not survive without Swainson's hawks, attending separately to their conservation would be a reasonable strategy for the system—and we would have identified a "keystone" species and a suitable subject for efforts to conserve the natural community. But if the hawks are no more critical to our principal objective than that nice patch of woods in the Darby watershed—if all we can say is that having the hawks is awfully nice and we wish we had more of them, or even if the hawks are an "indicator" species that confirms the presence of the community we mean to conserve— then we have defined a distraction, an obstacle to focus, and a step on the path that leads away from optimum conservation accomplishment. We ought to address ourselves to conserving the forest and grasslands community and presume that the hawk will come along with it. Indeed, if we make an indicator species the subject of special conservation efforts—if we artificially ensure its presence—it can no longer serve as an indicator. As nice as it would be if we could do it, conserving a community species by species is not often going to be a viable strategy.

It is not at all unnatural for conservation projects to drift off center as the Conservancy projects at Cosumnes and Big Darby did. There is, at every conservation site, so much work that is interesting and important to do. But it isn't all equally important, and it doesn't all contribute equally to the accomplishment of system objectives. Our ability to discriminate among the conservation opportunities that beckon is enhanced when objectives are stated with clarity and referenced frequently.

The insistence upon rigorous recognition of resource limitations is not supposed to drive us to conservation planning that balances the load so intricately that it's likely our ark will capsize if the weather doesn't behave exactly as we have predicted. In the Baraboo Hills of south central Wisconsin, for example, one of the Conservancy's system objectives is to conserve enough forest to sustain a "source" population of interior forest breeding birds. The objective of sending surplus birds from the Baraboo Hills to disjunct and widely scattered forest fragments too small to sustain them ("sinks," in the vernacular) sounds at first like the woods at Darby and the hawks of Cosumnes. But as a way to gauge accomplishment of what must be the principal objective—maintaining a secure population in a core conservation area of a set of rapidly declining species—the bird export standard seems to be a thoughtful way to measure our success and build in a margin for scientific or social error.

Indeed, prominent among the risks in the playing out of a focused and disciplined approach to conservation are the compromises that will be made. We will find it necessary to depart often from the ideal of a natural landscape in order to accommodate human and economic activity. In every such accommodation, we risk imposing fundamentally on the ecological processes we seek to preserve over the long run. There is a danger on the one hand that we won't be rigorous enough in avoiding hobby conservation; there is a danger on the other that we will succumb to the pressure we feel to plan for every bit of human use that our information about ecological limits tells us the system will allow. We don't, and probably never will, know enough about what the limits really are to responsibly plan that level of use.

*   *   *

Finding the right balance between the human use we must accommodate and the conservation we must achieve is a difficult task in the best of circumstances. We surely will not be able to successfully meet that challenge unless our system objectives are well and carefully defined. That means, to review the basic points, deciding what our objective is, considering its ecology, and determining what physical and biological processes are and are not important to maintaining it. When that task is well done, consistent reference to the result will help prevent the innumerable side excursions that every conservation journey will offer. The main trail is long enough.

My advice is to avoid defining a system of such scope and magnitude that it will require more than a human generation to produce a measurable conservation effect. It is seldom within our capacity to understand the way such large systems work well enough to create a satisfactory system description. That makes it difficult for us to develop effective strategies; that is in part why we expect that it will take so long to produce a measurable conservation effect on such a large system. Society has, after all, limited resources and limited will for conservation, and that will which does exist has to be reinforced with conservation success at intervals more frequent than a generation.

Society also has a big appetite that needs relatively little stimulation for activity that is incompatible with conservation. It is in the feeding of that appetite that most of the second "S" of the next chapter originates.

CHAPTER 8

Threats: Biological Stresses, Social and Economic Sources

If the system or systems to be conserved have been identified and described with clarity, the second and third steps of the planning discipline can be productively taken. The object in stress and source analysis is to specify the phenomena that threaten the health or integrity of the system(s) and identify and begin to understand the activities by which they are generated.

The Nature Conservancy's commitment to threat analysis was gradually built in the process of pursuing many projects without first considering threats. Conservation problems that arose in some of these projects might have been avoided or more easily dealt with had threats been considered at an earlier stage. Threat analysis has, as a result of that experience, been a part of the Conservancy's planning methodology since it launched a systematic effort to undertake landscape-scale conservation efforts. A small but really useful insight from the Conservancy's planning experience is that threat analysis produces the best results when it is separated into two distinct steps, with the physical or biological effect on the system—the "stress"—expressed first, and the causal agent—the "source," in the nomenclature of the planning discipline—named second.

Every biological system evolved and exists in the presence of a variety of natural stresses. Fire is one such natural stress. It has a dual nature, and while it is essential to the maintenance of some species and natural communities, it is a stress in and to them as well—stress and response are part of the nature of fire-adapted elements of biodiversity. It is accurate, therefore, to call a biologically essential fire a stress, and it is also correct to advise that, for planning purposes, fire and most natural stresses to systems are better thought of as aspects of the system than stresses. Indeed, maintaining or restoring natural stresses may be a conservation priority; their absence may require conservation action. When? As a general rule, if the system on which we are working is relatively healthy, and if we can afford to be patient because at least a few other examples of the system remain, we ought to restrain ourselves, remembering that natural cycles and the natural stresses that sometimes drive them have wisdom and rhythm beside which our own pale. By comparison with nature, we are short-lived. As a consequence, we're stuck with an impatience in our perspective for which we are just about incapable of sufficiently compensating. That doesn't mean we shouldn't try, though, and giving nature a little credit is part of trying.

We do need to recognize, though, that some of our own precipitous actions have interrupted or destroyed natural cycles. Some scientists have suggested that our meddling has had such far-reaching effects that we ought to give up the idea that any system can be maintained without our management attention. That is a frightening idea that leads to an untenable conclusion. It amounts to treating the world as we finally had to treat the condor. We aren't capable of it.

We should also be wary of the other extreme; some commentators have criticized management initiatives (prescribed burning, for example) that seem sensible given our understanding of the way a natural system works. The basic foundation for that criticism goes, a bit tautologically, like this: We are supposed to be saving natural systems, and a system that requires management intervention is no longer a natural system. When we provide management we have capitulated; we have finally robbed the system of its natural character and defeated our purpose. While the reputation we have made for ourselves as managers of nature is beyond redemption (we ought to tape the medical maxim *primum non nocere*—"first of all, do no harm"—to the dashboards of our four-wheel drives), the consequences of the acts of our species that are hostile or indifferent to nature make it impossible for us to withdraw from the cast and watch the remainder of the play from the balcony. Though we ought to take a conservative view when considering the event and type of intervention, we have completed an ascendancy that finds us capable of eliminating or deciding to save large numbers of the species and natural communities with which we share the planet. That capability brings with it a moral obligation. With every reason to be humble about our qualifications for doing so, we must fulfill that obligation. That means affirmatively managing natural areas when our best current understanding indicates the need to do so.

There are two particularly compelling instances for managing a natural stress, and the case for each emerges from the same general circumstance just cited: previous activity of humankind. Interventive management is clearly justified when a threatened natural system is among the last good examples of its type. This is a more common circumstance than might be imagined; there may, for example, be no North American tallgrass prairie system left that is of sufficient size and natural quality that all of its characteristic and ecologically important plants and animals can, without management, be maintained within it. In Indiana, where I did my first professional conservation work, about 3 million acres of wet and dry prairie greeted the first European visitors to what is now the northwestern part of the state. That prairie survived the first wave of settlement because its thick sod wouldn't yield to the first plows. But better plows arrived. The prairie gave way, leaving the few dozen patches of native prairie that now exist—amounting to a couple thousand acres in total—to remind those who recognize them of what was. None of them is, individually, bigger than a few hundred acres.

Dramatic, if not quite so ruthlessly efficient, replacement of native prairie by "amber waves of grain" took place across the midcontinent range of the tallgrass, wherever the soil would yield to a plow. If there remained a large North American prairie on deep soil and it were threatened with a serious stress that might be deemed natural, we surely could not let the stress run its course without a serious examination of the meaning of the extirpation of that for the continent's biodiversity.

The other easy case for intervention arises when we are challenged to conserve a system that is under stress from a "natural" condition that would have been diminished, made less frequent, or eliminated by ecological processes that have been disrupted. When scouring flood or wildfire has been removed from a system that was adapted to it, the system may become unbalanced and therefore unacceptably vulnerable to stresses that would, in other circumstances, be routinely absorbed.

* * *

For most conservationists, worrying primarily about natural stresses would be a luxury. Habitat degradation or destruction, rather than any natural stress, is what stares back when we look to see what's wrong with a system we want to conserve. Habitat has indeed been destroyed across most of the landscape of North America. For planning purposes, however, "habitat destruction" isn't a sufficiently useful description of a stress.

We destroy habitat by putting unnatural quantities of natural substances into it; by putting substances that don't appear in nature into nature; by setting organisms that belong in one place loose in another that is quite unprepared for them; by cutting up natural habitat into smaller and smaller pieces that become increasingly isolated; and by replacing native vegetation with exotic combinations of our own design. Each of these habitat-destroying activities has distinctive effects on systems we're concerned about, and the more specifically we can describe those effects—the stresses—the easier our next tasks—identifying sources and developing strategies—become.

Biodiversity values are threatened by an extraordinary array of stresses that can't be said to be natural. To select only a few illustrative pairings of systems and the stresses that threaten them, we might begin with the purple loosestrife that has escaped cultivation and found some important North American marshes to its liking. It is an exotic and invasive wetland species that physically displaces native vegetation. The small amount of native Hawaiian forest that hasn't been cleared is subject to physical destruction, introduction of disease, and accidental cultivation of exotic weeds by feral hogs. The absence of scouring floods is a stress to riparian habitat along the San Pedro River in southeastern Arizona and to the midriver sandbar habitat of sandhill cranes on the Platte River in Nebraska. In the upper Colorado River system, development of a conceptual model clearly revealed the importance of channel migration and thereby highlighted bank stabilization as a stress.

Sometimes the introduction of a basic stress into a system results in the generation of other biological and physical stresses. Nutrient enrichment that stresses a lake or marine system, for example, may result in an algae bloom. As the algae die and decompose, the water may show a reduction in dissolved oxygen, which may result in an altered balance of fish populations—which may itself have some stressful effects. A thoughtful stress analysis could begin at any of those points of stress, or in some symptom of one of them, and identify link by link each of the related stresses in the chain. That can be a useful thing to do because there may be strategies that mitigate the effect of the chain of stresses on the system while sparing us the necessity of directly confronting the economic activity that generates the first link, the nutrients.

Stresses should be stated in neutral and objective terms. We are highly partisan about the future we want to secure for the system. We often think, when we start working on a system, that we see how it can be made right. It is, therefore, especially important that our planning methods facilitate analysis designed to illuminate the subtle and indirect paths that will lead to the destination we seek as well as the obvious paths. That is one of the purposes of the planning discipline in general, and that is the aim of this advice on stating the stress.

One Conservancy team tried, in the midst of an extraordinary building boom that echoed all around it, to identify the stresses that threatened a nearby coral reef system of extraordinary biological and economic importance. The team initially concluded that the stress threatening the aquatic system originated with onshore residential and commercial development. The source now appears to have been more remote in origin. It is impossible not to suspect, in retrospect, that the planners' analysis and conclusions were affected by the development frenzy they witnessed daily. Whether they were or were not the story holds a lesson for all stress analyses: most of us can put to good use a discipline for managing our predilections.

*   *   *

Along the Roanoke River in northeastern North Carolina, the Conservancy's system conservation objective is an exceptional assemblage of bottomland hardwood forest communities. The structure and composition of these communities are changing. Certain tree species seem to be producing far fewer seedlings, and the population of other tree species in the Roanoke floodplain is increasing. The relative reproductive success of tree species along the Roanoke has changed. The project team juxtaposed that symptom with the knowledge that the flooding regime of the lower Roanoke is not following its historical pattern and was able to perceive a primary stress: the hydrological regime has been altered. There's too much water too early, average annual peak flood levels have been damped, and the floodwater that does come down into the lower Roanoke remains for a longer period of time than it used to. You don't have to look very far upstream to discover the source of this stress. The Roanoke has been dammed, and the regulated releases from the dam are not a close imitation of the natural flows of the river.

In the Ozarks, where the Conservancy is trying to conserve and restore a different set of forest communities, the symptoms are similar. The species composition and structure of the communities are changing. Those characteristic trees that were not eliminated as important community elements years ago through overharvesting are having difficulty with germination, while other tree species, historically less common in the highlands that straddle the Missouri-Arkansas border, have recently been doing unusually well. The explanation for the change in the natural community wasn't immediately obvious, but the problem for study was well isolated, and the stage was set for the absence of fire to eventually become conspicuous as a primary contemporary stress.

*   *   *

Besides specifically describing stresses that are clearly present, we will often need to decide how to respond to foreseeable stresses that are not currently affecting the system. Likelihood, timing, scope, effect, severity, and predictability are issues that have to be considered in a system when oil or chemical spills seem to threaten. Global climate change presents a similar analytical problem.

The Conservancy became interested, for example, in Delaware Bay, primarily because of its importance as a migratory stopover for shorebirds. In excess of 52 percent of the Western Hemisphere's breeding population of red knots and 31 percent of its sanderlings gorge on horseshoe crab eggs there on their way to breeding grounds each spring; significant numbers of raptors, passerines, and shorebirds appear in the fall as well.

The bay is also host to a large amount of ship traffic, including about a thousand oil-bearing ships carrying, in total, about a billion and a quarter barrels every year. The likelihood of an oil spill is predictable; it was recently

Figure 8-1. Red Knots on Delaware Bay. © Gary Meszaros

reported that the odds are better than nine to one that there will be a spill of 1,000 barrels during the next ten years. For a 10,000-barrel spill, the odds are better (or from a conservation perspective, worse) than six to one. Spills outside of port are thought to be nearly as likely. If the magic number comes up in the bay in the spring, the effects could be disastrous; at other seasons, the damage would depend on the effect the spill had on elements of the system other than shorebirds. What, for example, would a spill do to horseshoe crabs; what to the organisms on which they feed?

It is the answers to those questions that provide the first level of analysis in determining whether to list the effects of a spill as a stress. The first step is to go back to the descriptions and conceptual models of the systems we want to conserve. We then superimpose on those models the potentially serious but uncertain stress generated from a source such as an oil spill or greenhouse gas pollution. If the effect would be serious, we look to the likelihood of the occurrence of the stress, and to the timing. All but the least likely and most remote in time ought to be listed, even though there is an imperative associated with listing that prevails on us to determine sources and consider strategies. We can modulate our pursuit of source and strategy of a listed stress in accordance with our assessment of its importance, but the decision to list is nevertheless an important one—the character of a conservation project is profoundly affected by the stresses that must be addressed and the sources that are responsible for them. We surely must accept the real and important challenges, but no project needs phantom issues.

On Delaware Bay, the probability that stress from an oil spill would occur was known to be high and the potential impact substantial. The project team therefore decided to list oil spills as a stress, even though by so doing they signed on to do a lot of work to prevent or mitigate a problem that might never come to pass. A preview of the work: the risk of a spill could be reduced substantially if, to pick a potential strategy as an example without having researched its implications, double-hulled tankers were used in the bay. This is not a position to advocate casually because it probably comes attached to some formidable adversaries. It might be productive, before adopting it, to find out a little more about potential sources—to have a look, for example, at oil spill statistics. Do vessels flagged by a particular set of countries have markedly more spills than the rest? Does the risk of a spill become acceptably remote if permission to haul oil in the bay is granted only to the ships of the nations with the best safety records? Who decides who can haul oil in the bay? The decision to list a stress brings with it the promise of an excursion into source and strategy.

Climate change is another commonly encountered member of a class of stresses that are likely to happen, but indeterminate in time, scope, and effect. Recommending an appropriate response to such stresses is problematic. Though it appears quite likely that societal release of greenhouse gases will be the source of a physical stress—changes in atmospheric composition—that generates a consequent physical stress—climate change—our ability to relate such a global phenomenon to noncoastal systems is limited because our understanding of the precise effect of global climate change on most local systems is so limited; most will surely be warmer, some wetter, some dryer. Until better models are developed, we probably can't describe the local stress well enough in most inland systems to justify even the citation of climate change as a stress for planning purposes.

The balance is on the other side for a coastal system. Current information and models strongly suggest that climate change will result in clear and predictable problems—for example, physical destruction from increased wave action—for coastal systems. The likely problems ought, therefore, to be acknowledged as stresses. By doing so, we help illuminate at this early point of the process paths toward solutions that would involve, for example, mitigating locally the physical beating the system would take. The obvious and broader challenge of addressing global climate change needs no such highlighting.

To foreshadow source and strategy discussions that will emerge later, there are things that can and must be done, and are being done, to avoid or blunt the effects of increased concentrations of carbon dioxide, methane, and other gases in the planet's atmosphere, but they aren't project specific or project scale. To recall the earlier discussion about choosing a system as a conservation objective, we get only half of what we want if we choose to work on a system principally because it provides entrée to a larger problem. We certainly want to work on systems that exhibit localized threats that are common to many systems; that is one way to get the leverage we need from the expenditure of our limited resources. But it isn't likely that we will find ways to get strategic value from a system that merely exhibits an unremarkable local manifestation of a global stress.

The planning discipline for an ecosystem approach to conservation is, almost by its terms, a site-centered approach to conservation, and there are certain weaknesses as well as strengths inherent in that approach. Site-based conservation can be an important strategy for addressing large-scale conservation problems if lessons learned can be disseminated so effectively that they stimulate other locally based conservation activity. Working at even a very large number of local sites is not, however, by nature an optimum approach for addressing stresses that are highly dispersed in generation and effect (though it may, in the end, work as well as other techniques even for that kind of stress). Poorly executed, however, this approach is vulnerable to a tendency to deal with exportable stresses by exporting them—like dealing with too much smoke by building a tall stack. That vulnerability can be managed by making a commitment not to solve problems in that way, or it can be managed by legal mandate: treaty and regulation that make it illegal to solve local problems by contributing to the creation of problems elsewhere.

*   *   *

The second part of threat analysis, and the third step of the planning discipline, to reiterate, is the tracing of stresses to their sources. The stress of physical destruction of coastal habitat can, for example, be traced to increased wave action, which is caused by sea level rise, which can be traced to warming of polar regions, which is a manifestation of the greenhouse effect, which can be traced to increased concentrations of CO_2 and other gases in the atmosphere, the source of which is burning of fossil fuels. The source of that excess fire and the ultimate source of that coastal destruction—indeed, the ultimate source of most stresses—is human population of twentieth century magnitude. How far up the chain is it productive to trace stresses to their sources?

It may be useful here to think as a lawyer would and trace causation to its "proximate" source. In tort law, the proximate cause of an accident defines the limit of legal responsibility. There exists no precise formula by which to determine in advance whether an entity that can be linked to an accident is close enough to be held to have been proximate. It is, literally, a judgment call at the boundary, part practical and part policy (with an occasional, rarely seen, overlay of common sense).

So it is with the definition of source. For planning discipline purposes, sources are the proximate causes of ecological stresses. The purple loosestrife that stresses a North American wetland may have as its source seeds that escaped from ornamental plantings. On the Platte River, the absence of scouring floods can be traced to the operation of dams and the drawing down of surface water and groundwater. Feral hogs became a potential source of stress in Hawaiian forests with the introduction of wide-ranging European swine to the islands (the pigs of the Polynesians apparently didn't range far from the villages of the people who brought them to the islands). The potential was actuated when forests were diminished and fragmented and non-native vegetation began to dominate the cleared land. Both the hogs and the exotic vegetation got loose in Hawaii because tourism and trade has not been conducted so as to avoid introduction of alien species. In an environment so isolated ecologically, intentional or accidental importation of alien species at a rate orders of magnitude higher than natural is a deadly serious source of stress.

Each source of stress just cited is rooted in the demands of a growing population. The conservation sensibilities and capabilities we have now developed might well have been up to the job defined by a population of the size the planet hosted two hundred years ago. But there has been a sevenfold increase in human population since 1800. There isn't any useful alternative to tacitly conceding certain battles to the wants and needs of some 6 billion people. The demands of an unprecedented number of humans on the earth, collectively, is a major driver of system stress; at some point, it becomes the source of stresses for which there is no strategic solution but better success at family planning. But that point will vary with the environmental consciousness, as well as the technology, of the population. Identifying global population growth as a source of stress for purposes of project planning probably won't be a productive exercise; it's usually not a proximate source of system stress. We clearly need environmentalists to apply their resources to population as a global issue, but system conservationists ought to play a different role. Their resources should be applied to population issues only when they are likely to produce important results soon—where there is a specific and direct local manifestation of the generic and global source.

A similar kind of reasoning can guide us with respect to other global sources. It would be a good thing if we as a world society used less and reused more wood and paper products, but addressing our strategic efforts to that broad-based problem will not produce timely results for a specific threatened forest ecosystem. Solutions that will work quickly enough to save the system will concede the existence of a societal norm that needs reforming but will deal with the local manifestation of the problem locally, with the objective of rendering the effect of the problematic practice uncharacteristically mild in the project area. It is, again, useful to describe sources, even sources that exhibit global expressions, in local terms; for a stress that might be expressed as "change in forest community composition and structure," the source might be described simply as "overharvest"; tracing the causal chain to "global demand for wood and paper" probably will not prove useful. If the source of stress in an Adirondacks forest system is overharvest, a responsive strategy might be to create a business opportunity that could generate comparable income with a lesser cut. That may mean, for example, working to pilot and prove out the economic viability of value-added forest product processing as a more attractive business than pulp production in a system that will always be pushed to overharvest if it must compete with southern forests at pulp production.

To cite a different example, The Nature Conservancy didn't name the global phenomenon of oxidation of carbon-based compounds as a source in its Florida Keys plan. The project team is not going to address global climate change in any way at this stage of the project; team members have their hands full dealing with imminent threats originating right in Florida. The bleached appearance and reduced coverage of live coral made it clear that the primary system objective, the reef, was undergoing biological stress (as well as less important but more manageable localized physical stress from the activity of divers and the anchors of the boats that brought them). It was, however, a good while before the team could make enough sense of the symptoms of stress to cite a source. There were nearly as many theories about the nature of the stress as there were scientists who studied the matter. The Conservancy chose to begin to work on the best available consensus as to threats while investing, at the same time, in the science that would support a second-generation hypothesis regarding stresses and sources.

The first conservation work was directed toward the initial consensus choice for a source: development in the Keys. It seemed to the Conservancy and the consultants it engaged that the increased level of building was resulting in nutrient enrichment from inadequate sewage treatment, which must be harmful—a stress—to the reef. That hypothesis hasn't been proven wrong (though nutrient enrichment may be primarily a near-shore problem in the Keys), but a second-generation hypothesis has since been developed, and it suggests that there are other, different sources that ought to be considered of greater immediate concern. The likeliest explanation for the stress the reef is experiencing is that temperature and salinity gradients have changed in the water flowing across the reef from Florida Bay. The bay, in turn, is fed by the sheet of water flowing across the Everglades, which is reduced in volume and quality by the sugar production and other agricultural activities that share the south of Florida with the glades. The focus of the conservation work has, obviously, been adjusted toward those sources.

Was the Conservancy wrong to initiate a conservation program designed to address Keys development as a source of nutrient enrichment? A principal theme of the planning discipline is recognition of and planning for resource limitations. The advice to the project teams at Big Darby and Cosumnes was to define their systems narrowly. It seemed important in those projects to avoid devoting resources to the pursuit of secondary priorities, precisely because the conservation objectives were secondary and would divert limited money and attention from primary objectives.

Acuity is critical. The first programs of the Keys project were directed toward a source that now appears to be secondary. But the decisions of the project team were sound—and the team was observing an earlier-stated rule that is even more fundamental: act. Get informed, recognize the risks, try to avoid doing harm, but act. It is only by acting that we are afforded the opportunity to react. The decisions the Keys project team made also embodied another Conservancy value already stated: act on the best *available* science.

The planning discipline and the conservation it guides are interactive processes. The dialogue cannot commence until someone makes a statement. The problems we face are too urgent to allow us to be still until we are certain the first statement we make will be eloquent. The best chance we have to eventually make it so is to begin talking, begin acting, gauge the reaction, and make appropriate adjustment. In the Keys project, the Conservancy did fine. It both initiated action and invested in development of a second-generation hypothesis as to the critical stress and source. When a second-generation hypothesis emerged, a second-generation set of strategies was developed.

There is a point at which an initial hypothesis will not be sufficiently grounded to justify action. A Conservancy vice president once said that haphazard conservation is as bad as haphazard development. It isn't, of course. But the reasoning behind the aphorism is interesting: the idea is that if we insist on conserving areas that aren't priorities, we have at the least left open to development the areas that ought to be the first to be conserved. We may indeed, through a process of displacement, have effectively encouraged development to proceed in the very locations that ought to have been conserved. An equally compelling problem with ad hoc conservation is that in pursuing such an approach we will expend conservation's limited resources and society's limited tolerance on areas that don't offer the greatest return possible on the expenditure.

The call made here for action is not, in other words, meant to suggest that any well-intentioned effort will somehow produce a good result. It isn't likely that ill-considered movement will generate a reaction that magically makes it possible to proceed from a blind first step to a well-directed second. A reasonable basis for a choice of direction—a hypothesis about stress and source—is required or, at the least, enough knowledge to support a first approximation of what isn't known. The Conservancy satisfied those requirements when it began to take action in the Keys.

*　*　*

Each step of the planning discipline prepares the way for the next. In the same way that a good definition of system makes stresses identifiable and well-defined stresses point clearly to the source, a good understanding of source will permit a better initial statement of strategy. Source, like stress, ought to be identified in very specific terms that are as neutral as possible. Thus, in the Big Darby, the source of the siltation stress isn't heedless farmers. The source is conventional tillage and, particularly, conventional tillage that is extended to creek edge. Along the Roanoke, the source is the manner in which the upstream dam is operated and downstream releases are regulated.

If the distinctions seem artificial and the means of expression calculated, they are. The discipline exists to enhance the possibility that conservation planners can develop strategies that will work. Conservationists usually don't need help in summoning sufficient passion for their work. Cultivating a managed and somewhat clinical planning language is a modest attempt to add an overlay of objectivity that will help us productively direct the energy that our passion generates.

There will remain a need, to make the point again, for conservationists who take it as their role to confront society with, say, the Roanoke dam itself or with any of an endless variety of ill-considered sources of stress as the issue. If an organization that will mount such a challenge does not already exist in a project area, the project team might even consider helping to create one. But, to recall another point that bears repeating here, when it comes to getting a source to change a behavior so that it alleviates the stress, the shortest distance between two points is frequently not a straight line. And the need for change is often so urgent that finding the best path is critical.

*　*　*

Successful system conservation requires that we think about addressing sources of stress in ways that require only measured and subtle changes in powerful social and economic institutions. Ideally, these changes will be the first on a path that leads to more fundamental changes. But good strategies can move a source to relieve the system from critical stresses whether other changes occur or not.

When we believe we know the source of problems in the systems we care about, we tend to want to get right at those sources of the problem and take them on directly. Some important environmental gains have been made by following that approach. But it has failed more frequently than it has succeeded. Confrontation as an approach produces news, but change doesn't always follow. The watercourse way of Taoist tradition is a metaphor for an approach that may work better: the sources tend to be rocks, and it's easier to find ways to flow over or around them than to break through them.

That advice will be further developed in the chapter on strategy, which follows the next chapter's discussion of sorting through systems and threats to set priorities.

CHAPTER 9

Priorities for Action: Sorting It Out

It isn't too difficult to describe the process of identifying systems, stresses, and sources and, in doing so, to cite examples that illustrate both successful and problematic efforts to use the planning discipline. But understanding systems, stresses, and sources will be more challenging in a project that is not a particularly clear example of a success or a problem in applying one of those S's. In most projects there will be numerous interlocking systems, a variety of stresses, and diverse sources. Project team members will have an intuitive feel for how to sort these out, and if their intuition is good, that may be all that they need. Many Conservancy project teams have, on the other hand, found it useful to approach the challenge of sorting it out with the aid of a process for ranking the results obtained in the systems-stresses-sources analysis.

The objective of such a process is to determine which of the threats—the linked stresses and sources—are the most important to address with strategies. Which, in other words, are the most serious concerns with respect to the highest-priority elements of the natural system? The process also serves to highlight stresses and sources about which we know so little that we cannot make good judgments with respect to their effect or importance, and to which, therefore, additional research should be directed.

Some of The Nature Conservancy's project teams have suggested that they've been able to make better sense of the priority-setting questions by creating a table in which the systems present are listed along one axis in order of priority and the sources are listed along the other in order of overall effect. The table is a snapshot, and the subject will look different over time as strategies are applied to alleviate the threats to the systems.

To prepare the data for making such a table (or some other display for sorting priorities), the first step is to list the systems that have been identified as priorities and establish an order of priority among them. This is particularly important if the project's setting—the physical area in which the work is to be done—was chosen geographically rather than biologically. Even when the choice was made for biological reasons there will often be more than one system that needs attention, and distinctly different conservation approaches may be needed. At Virginia's Eastern Shore, for example, the colonial nesting birds, a conservation priority whose characteristics and requirements can be set forth as a system, have quite different conservation requirements from the migratory songbirds, for which a different part of the land in the project area is important. Acknowledging that is not to belie the point made in chapter 7; choices still need to be made and priorities set. But even when a rigorous process for doing so is followed, there will often remain more than one system on which to work. In the Conservancy, degree of endangerment provides a basic guide for setting priorities among priority systems. Other conservationists must make that determination with reference to their own fundamental purposes.

After listing and ranking the system or systems to be conserved, the project team can turn to the stresses and list them system by system. The nature of stress identification in the planning discipline is that the listed stresses will differ in the intensity, frequency, and duration of their impact and, perhaps, in the likelihood that they will occur. To facilitate a sorting-out process, the stresses then ought to be ranked in order of priority. A qualitative ranking hierarchy ranging from low through medium and high to very high has proved adequate in the Conservancy's experience.

Factors that the Conservancy's planners consider in assigning a rank to a stress include the severity and scope of the damage that the stress will cause: is it likely to destroy the system or seriously degrade it, or will the stress only impair its operation? Is it pervasive or localized? Other things being equal, pervasive stresses are more serious than patchy or localized stresses.

The state of our ability to understand the workings of ecosystems is such that we will often recognize symptoms of stress without being certain of their cause. The Conservancy knew, to recall an earlier example, of damage to the reef system of the Florida Keys but could not for several years say with confidence what stress actually seemed to be causing the damage. For purposes of planning and taking conservation action, a stress should be ranked as a more serious one if its impact is thought to be certain than if there are doubts about its impact.

Each of the enumerated stresses will have emanated from one or more sources that themselves ought to be ranked. In assessing the relative seriousness of a source, it is important to consider both the current and future contributions it will make to listed stresses. Current contribution moves a source up on the list of priorities, though a source with irreversible effects or a source that works as a ratchet—it moves with power in only one direction—may ultimately be adjudged more serious. New residential or commercial development that is not planned to be compatible with the local environment, therefore, may end up with a higher-priority rank than farming practices even if the development threat is around the corner and the farming impacts are being felt today. It is usually even more difficult to alleviate stresses arising from building and development than it is to interest farmers in making changes in the way they go about their business—and the stresses associated with intensive development tend to be among the most acute that conservationists face.

Obviously, if the risk of stresses emanating from development is remote, either because development is a remote possibility or because the development in question is not certain to be a source of stress to the identified system, that source would receive a correspondingly lower rank.

Table 9-1 is based on an analysis of stress carried out by a conservation team working in the Baraboo Hills of central Wisconsin. There, residential development and logging as currently conducted emerge as sources of stress that require priority attention. Quarrying, by contrast, appears to be a less important concern.

The same source may contribute to several stresses; Everglades water management, for example, is thought to contribute to four different stresses in the Florida Keys. It has a low rank with respect to nutrient enrichment but a high rank when it comes to salinity change.

When all stresses and sources have been listed and ranked, the project team can begin to make sense of the task of assessing overall priorities for the purpose of creating strategies. A source that appears many times, even in connection with important stresses, still may not be addressed with the first round of strategies if there are other sources that have a very high impact on very highly ranked stresses in high-priority systems. Some Conservancy planners have found it useful to assign numerical values to the component rankings and to consider overall priority by performing the arithmetic and comparing the results. The apparent precision of the numbers, however, can be no better than the necessarily subjective system used for assigning them.

The ultimate result we are seeking with the sorting-out process is guidance as to where to direct the creative process of designing strategies that will effectively address stresses. Before discussing strategy, one more note. Time after time, having listened to competent and thorough presentations of project plans for system conservation, one of the Conservancy's most experienced reviewers would ask project teams to identify the "killer" threat. The teams' reactions were generally along the lines of "We just told you about a variety of systems and told you about the priority we assigned to the stresses and sources we have identified." The reviewer wanted something more. He wanted to know that the project team could describe in a few words the most difficult challenge the project would face, and he wanted to know that the

Table 9-1   Threats analysis for the natural community mosaic of the Baraboo Hills, central Wisconsin

SOURCE	STRESS — Overall Rank	Habitat Fragmentation	Habitat Loss	Displacement of Native Species by Exotics	Habitat Conversion	Degradation of Water Quality	Microclimate/Microhabitat Alteration
Residential development	VH	VH	VH	H	M	M	L
Detrimental logging practices	VH	VH	M	VH	VH	H	VH
Fire suppression	H	L	L	VH	VH	L	L
Roads and corridors	H	VH	M	M	L	M	L
Agricultural practices	H	L	L	L	VH	H	L
Recreation	M	M	M	H	L	L	L
Silvicultural practices	M	H	L	L	L	L	L
Quarrying	L	M	M	L	L	L	L
Agricultural conversion	L	L	L	L	L	L	L

Note: L = low; M = medium; H = high; VH = very high.

team was planning to deal with that challenge. It was surprising how often the answer revealed something that went beyond the presentation and the planning material.

When a joint Conservancy-ANCON (a leading Panamanian conservation organization) project team presented a plan for conservation of the Darien to a group of reviewers, for example, it described compellingly that wild and relatively remote area at Panama's border with Colombia. The team dealt with stresses associated with timbering, in-migration, struggling indigenous peoples, and the potential interpolation of the Pan-American Highway through the "Darien Gap." In the discussion that followed the presentation, though, the highway emerged as the one threat that could accelerate all others, and the only strategy documented was opposition. But do a fine tropical forest and environmentalist opposition win against the only discontinuity in a road that runs from Canada to Argentina? Maybe, with the assistance of animal science professionals who see the gap as a barrier to transmission of some diseases of livestock that are, to date, largely contained in South America. But good long-term planning probably requires the development of an ecologically tolerable alternative. The planners knew that, but their consciousness of its fundamental importance receded as they considered and documented the full range of threats. Newer iterations of their planning processes make responding to the road a central priority.

It is often the killer threat that really tests a project team's strategic thinking, and it will, as well, often be the addressing of that threat that contributes most to meeting the conservation challenges that face the world.

CHAPTER 10

Living with Stresses and Working with Sources: Conservation Strategy

Strategy should be the climax of a plan for an ecosystem approach to conservation. The careful discipline that attends system, stress, and source definition can be loosened when it comes to developing strategy; the more creative and powerful the strategy, the more compelling the climax.

Strategies for conservation nearly always require that people change something they are doing, are planning to do, or might someday want to do. Good strategy incorporates or at least acknowledges the things people hold dear as we ask them to change.

It is people, as well, who must ultimately defend the natural systems we want to conserve. A proper foundation for effective conservation strategies, therefore, will include knowledge of the social and economic context in which relevant human communities exist. As a movement, conservation has been weak in this area. Our collective experience is weighted toward influencing national legislation and policy rather than working with local communities. The Nature Conservancy's particular approach has required that it work with thousands of local landowners, and as a landowner itself it has attended when it could to the responsibilities of being a good neighbor. But it too has traditionally paid little attention to local communities—it was busy getting after the next land deal. Recently, Conservancy people have begun to assimilate into their work habits the imperatives of the ecosystem approach to conservation. An organizational sense that human communities are in a position to conserve or disrupt those processes has begun to emerge, and working with human communities has consequently assumed a place of importance.

Which communities are relevant for biodiversity? Community isn't a concept that admits of immutable definition. The boundaries of the relevant community for a conservation project, if they were ever clear, are clear no longer. There are communities of year-round residents, communities that include summer or winter residents, economic communities, electronic communities, and communities of every conceivable sort organized by common interest.

In some places, conservationists who would not otherwise be considered members of the community in which they are working find common ground—a sort of community in itself—with community members in a love or appreciation for the land; many people seem to have a kind of residual memory that the land is the foundation on which security and prosperity rests. Among people living in and around areas of sufficient natural quality to be conservation priorities, that memory is often even sharper.

If we look hard enough we can find evidence of an even deeper, inborn love of nature, sometimes called "biophilia," in almost any community. The coal company executive has duck prints on the wall, has tropical plants in the front office, and lives in a neighborhood of upscale homes accented by the mature trees the builders carefully left to enhance the market appeal of the development. Knowing of that bond may avail us of nothing more than encouragement to keep trying to come up with a strategy that works—but maybe that is enough.

For land conservation purposes, communities of residents, both seasonal and year-round, are of the greatest importance. They will feel the consequences of conservation failure most acutely, and they are most often in a position to extend or withhold controls in respect to land uses. But an analysis must be made in every project to determine which communities are present and which are in a position to influence the conservation outcome we seek.

As we try to understand what people mean to the system we want to conserve, the opening question is "Who lives on land that affects the system?" Then, "How many are there, is the number increasing or decreasing, and how fast; what are the age, educational, and income profiles and how are they changing; what do residents believe are the area's principal problems and aspirations; how, and in what forums, do they and their neighbors make decisions when they have to make decisions together; what is the nature of land ownership and tenure; what are the major industries and employers, and how is the land used?" Understanding the relationship between employment and land use is often particularly critical as background for developing strategy.

Assembling this kind of information about the community is due preparation for initiating a dialogue with its members or representatives about the function and future of the natural system in which they live. When should that dialogue begin? Many conservationists and most community activists argue that it should begin as early as possible. A process that is opened up early and remains transparent throughout is always good form and, if brilliantly facilitated or unusually blessed with luck, can lead to good results.

In all cases, a "no surprises" rule is essential. That is, the conservation project's representatives must commit themselves to informing community representatives in advance of any major transactions that will affect them. There is a difference, though, between the "no surprises" rule and a process that is open early and always. An honest and inclusive working relationship with the community is critical. But I have concluded, from the Conservancy's experience, that a little more, rather than a little less, preparation before asking for the involvement of a broadly representative part of the community will render a conservationist better able to provide assistance to the community and more likely to achieve conservation goals. It's good, for example, to know that transfer payments are the largest source of income to a southern Illinois community we want to work with on the Cache River. It's even better to be able to identify the diverse sources of such payments, to have an informed perception of the relationship of transfer payments to the systems and stresses that have been identified, and to have considered the manner in which strategies that might be considered are affected by those facts. That kind of preparation is required in advance of making a preliminary commitment to any strategy. That kind of information will reveal paths we can take to arrive at strategies that are worthy of the name because they are formed in the light of the circumstances of the relevant community.

Strategy ideas are the most valuable currency conservationists have to expend in broader community processes; they will be influenced, changed, and supplemented by the processes, but in my view, it is better to come to the processes with idea currency instead of empty pockets.

Good strategy ideas are developed as a problem-solving exercise. The threat analysis discussed in chapter 8 will reveal social or economic activities that produce ecological stress. Those activities will have been initiated to satisfy human needs or wants. The problem to be solved is the reasonable satisfaction or displacement of the stress-producing needs or wants by means that do not generate the stress.

If the earlier-discussed dam on the Roanoke is a given, is there a way to operate it to more closely mimic natural hydrological cycles downstream? Is there a way to minimize the risk of oil spills in Delaware Bay during the critical seasons that doesn't line up the shipping interests as a strong and unified force in opposition? Could we slow harvest in a forest system if value-added production replaced pulp, and what kind of investment in equipment and training and development of new markets would make value-added lumber processing a viable alternative? Are there alternative ways to satisfy the needs that drive the diversion and pollution of freshwater in south Florida, and can the needs of the sport and commercial fishermen who use Florida Bay help galvanize interest in exploring any such ways?

Asking such questions at the Big Darby (introduced in chapter 7) produced a number of answers that bear pursuit as strategies. There, silt is a major stress in a system that supports fishes and mussels. Tillage associated with conventional farming practices for corn and soybean production is a source. Conservation tillage would help because reduced plowing and cultivation would leave less soil exposed to erosion. But change implies risk, and the risk of adverse effects on costs or yields because of a change in tillage practices has been enough to keep conventional tillage the norm in the watershed. The project team needed another conservation tool.

Team members began to wonder whether there were agricultural products that demanded less of the system. Hay seemed to offer competitive economic returns. They found, however, that the hay market was thin and undependable by comparison with the well-developed markets that had formed around corn and beans. Also, drying the hay to suit likely buyers was a problem; Ohio is relatively rainy compared with other hay-producing areas. The team is now developing market information with the hope of addressing the first problem and experimenting with solar dryers as an answer to the second.

The effort is worth the trouble because hay production would dramatically reduce plowing and thus siltation. Also, hay will thrive on significantly lighter loads of farm chemicals than corn and soybeans have usually demanded. Most important, production of hay wouldn't require major changes in lifestyle and work habit for the people the Conservancy hopes to persuade to plant it. Finally, a related advantage is that hay production occupies land and to that degree contributes to the stabilization of use and ownership of land. If farming ceases to occupy land in the Big Darby and farmers sell out, the buyers will be suburban developers. A new, different, and even more complex set of stresses will replace silt and farming as conservation concerns. Facilitating streamside hay production on existing farms has, in other words, the attributes of an excellent strategy. The farming community has shown enough interest in the idea to justify development of a detailed business plan for a hay enterprise.

A sound way to begin to think about natural resource–based production as a conservation tool is to consider broadly what items of economic value the region is particularly well suited to generate and what markets there are for those items—a kind of comparative economic advantage analysis, with a more-than-usually focused ecological filter. It may be useful to envision three intersecting spheres of interest, one containing the products landholders can and will produce, one containing products compatible with maintenance of a healthy system, and one containing products that can be sold at prices that return a profit or, better, a premium. Conservation organizations may be able to exert some influence on the third sphere with licensing or certification programs. The products to investigate for conservation production are, obviously, found at the intersection (figure 10-1).

Conservation production is an advanced strategy that is part of broader plans the Conservancy has made to protect the ecological systems of Virginia's Eastern Shore. Those plans have been developed over many drafts, revisions, and phases. The evolution in strategies is instructive.

The work on the shore had its origins in one of the crown jewels of the Conservancy's nature preserve system, a string of barrier islands that stretch from Assateague, at the Maryland-Virginia border, in an almost unbroken

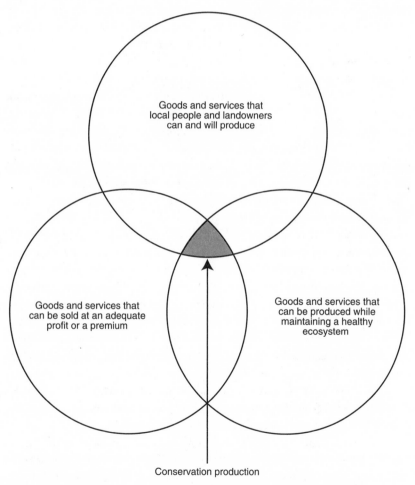

Figure 10-1. Natural resource–based production as a tool for conservation.

chain to the southern tip of the peninsula. There isn't a comparable expanse of wilderness off the Atlantic or Gulf coasts. Much of the core protected area is pristine because of the natural migration of the sandy barrier islands themselves. As the nineteenth century was coming to a close, many island buildings were moved by barge to the mainland. What was left was soon underwater. That, of course, was before builders and engineers developed enough tools for stabilizing a barrier island that a lender could reasonably hope somebody would keep such a site in one place long enough to amortize a development loan. It was little more than luck that placed the first modern waves of barrier island development somewhere other than the barrier islands of Virginia's Eastern Shore. The Conservancy arrived on the shore in time to buy most of them before the development wave that would have swamped them. The ever-shifting sand of the islands is at the heart of the system the Conservancy is trying to protect. As TNC closed the last major deals for acquisition, it began to look forward to a slower pace, a long, peaceful job of stewarding a peerless nature preserve.

Stewardship, though, is seldom a peaceful job for long, and on the Eastern Shore it has never been less than dynamic. In the very defining of the system, the Conservancy was confronted with the incompleteness of its grand acquisition: though the islands were the heart of the system we cared about, the shallow bays that separated them from the mainland were its breath and blood. Those bays, in turn, were fed by the creeks and rivers and salt marshes of the eastern half of the narrow peninsula that is the Virginia Eastern Shore. The creeks and marshes had suffered some from runoff and siltation when commodity crop agriculture came to the shore in the 1960s, but they were still comparatively healthy.

The mainland of the Eastern Shore had for several reasons been spared the booming development that was occurring around Norfolk and Virginia Beach. Because the southern tip of the peninsula is across the mouth of the Chesapeake Bay from the Norfolk metropolitan area, for years, someone considering a commute from a residence on the Eastern Shore to a job in Norfolk faced a couple of long ferry rides every day. When a privately financed toll bridge–tunnel succeeded the ferry, the barrier to commuting became an eighteen-dollar daily toll. But by the early 1980s, developers were beginning to show an interest both in commercial building at the southern tip of the Eastern Shore and in residential building wherever the peninsula afforded a view of the water. On a spit of land that ranges from four to twenty miles in width, not a lot of land was immune (figure 10-2).

All this was of secondary interest until the Conservancy began to prepare to meet the conservation obligations of a system description that did not stop at the edge of the islands. The blueprint read: island systems and salt marsh depend on shallow bays depend on shore watercourses depend on

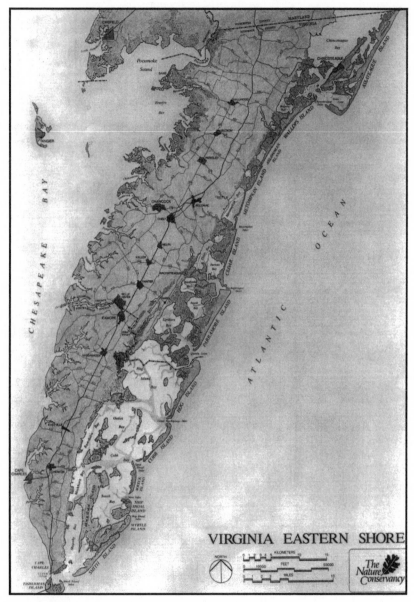

Figure 10-2. The Virginia Eastern Shore. Prepared for The Nature Conservancy by
Vladimir Gavrilovic, Paradigm Design, Reston, Virginia.

land use on the seaside of the peninsula. Commercial and residential development was beginning to present a challenge that had to be addressed. Seventy miles of seaside coast was in play, to say nothing of the Chesapeake Bay shore on the western side of the peninsula. The land acquisition approach that had taken the Conservancy so far was not going to be the sole conservation tool of the future. Further, had massive land acquisition been practical, its appropriateness would have been open to question. The mainland had been farmed by European Americans about as long as anywhere on the continent. The land was not nature preserve land, and it was not necessary to restore or reclaim it to preserve the region's natural values. Indeed, the natural character with which the Conservancy was concerned not only had survived human use of the mainland; it was an important part of the culture and identity of the watermen and the farmers who lived on the shore. Uses about as intense as those of the past three hundred years were compatible; dense residential, industrial, or commercial development represented a threat to the integrity of the water that is so central to the barrier island ecosystem of the Virginia Coast Reserve.

The first strategic innovation the Conservancy adopted was to initiate a new and more finely honed approach to land acquisition. Having observed the usual pattern of waterside development, the Conservancy acquired a package of land around the deepwater ports of the seaside that a group of conservation buyers had assembled. Those lands were acquired not to be nature preserves but to enable the Conservancy to become a force in the future of that seventy critical miles of Eastern Shore to the extent that the organization's limited resources allowed.

With those acquisitions, the Conservancy was clearly engaged on the mainland as well as on the islands that had first drawn it to the area. To anyone who was paying attention—and the Conservancy was now paying attention—it was becoming clear that the traditional farm life of mainland Eastern Shore Virginia was becoming more and more difficult to maintain. The value of the land was going up, and property taxes were following. Eastern Shore farmers had once grown mostly vegetables. But government farm programs eventually made corn and soybeans seem like a better bet. There isn't much to distinguish the corn and soybeans produced on the shore from those of any other region. And when the costs of one set of producers are higher, and the product they want to sell is the same, the eventual result, in a competitive system, is predictable.

TNC considered putting more resources into its efforts to seek conservation easements that would prohibit residential development on seaside farms. By promoting that kind of conveyance it hoped, farm by farm, to stabilize land use and preserve the general pattern of the landscape, maintaining in the process the rentable farmland that was essential to the viability of the working farmers who remained on the Shore. The easements it was using had been designed to contain affirmative obligations as well, such as "best practices" with respect to runoff and erosion control. The Conservancy is still seeking such easements, but the program never achieved the central importance originally envisioned. It's a time-consuming process to introduce, explain, and promote the benefits of conservation easements, and the incentives for landowners to give an easement or sell one for less than the land's development value are not so powerful that a farmer who isn't interested at the outset is likely to develop a strong interest fast.

That is a serious issue because without a critical mass of working farms, the broader farming community in a region starts to disintegrate, thus initiating a downward spiral that makes it ever harder for the farmers who remain to stay at it. The implement dealers, the service people, the knowledgeable money people all have roles to play in a farm economy. Their presence is nearly as important to an agricultural way of life as the farms themselves. Their absence would make what is already a hard way to make a living even harder.

There were farmers on the Eastern Shore who had been through that spiral elsewhere and hoped to avoid a second experience with it. The threat to a way of life that stretched four hundred years back was an issue at some level in the broader community, too. The Conservancy saw farming as a Shore land use that was far preferable to the alternatives. It was clear, nevertheless, that some farm owners were going to give up on the farming life and sell their land to the highest bidder. In response, the Conservancy added a third strategy. In its initial iteration, the strategy called for TNC to acquire lands that were going to be sold out of farming use. Costs would be recovered through resale, with restrictions designed to limit development to ecologically tolerable rather than legally permissible levels. As TNC became more thoughtful about the potential of that tool, it commissioned a model for limited development of such tracts. Buildings were to be located in the least obtrusive places on the land, and new buildings were designed to be consistent with the architectural styles that characterized the Eastern Shore. Some of the amenities of more common, more intrusive arrangements— water access, for example—were to be supplied, according to the model, by the creation of common areas. Retention of rentable farm fields—not so incidentally supporting the goal of preserving the farm community—seemed, surprisingly enough, to enhance the marketability of the building lots rather than detract from it.

While all this evolution was under way, TNC was getting to know the people and leaders of the community. Importantly, the converse was also true. Because of its aggressive and impressive efforts to secure the barrier islands from development, the Conservancy had achieved notoriety in the

Figure 10-3. Life on the Eastern Shore. © Curtis J. Badger

community years before. Not everyone was favorably impressed. Many viewed the Washington-based conservation organization, which had, overnight by Shore standards, become the largest landowner in the community, with suspicion and fear. In the early 1980s, the Conservancy hired a new manager for its Shore properties, a quiet, prematurely gray Georgian named John Hall.

Hall was a successful nature photographer, and he brought the eye for detail, the patience, and the timing that accompanied that expertise to his new job. The rural background, the naturalist's understanding of the salt marsh, and the soft southern accent didn't hurt. Over the next few years, the Conservancy became a lot less the invader from D.C. and a lot more the solid countenance, the patient manner, and the gentle but firm resolve of John Hall.

A few years after Hall arrived, the Conservancy began to provide assistance to a local civic organization that was trying to design and secure passage of an improved zoning ordinance. Central to that assistance was finding some funding for public opinion research—a poll. The poll revealed that a large majority of respondents opposed zoning. At the same time, though, similarly large majorities wanted "local control of growth" and to "preserve the character and quality" of the area. An ordinance was designed that would confer some of that ability to control growth and help preserve the area's character. The measure that was enacted was temporary, but it was a big step forward. If the goal of improved zoning was a rose, the new development guidelines smelled nearly as sweet.

Over the course of a couple of years, the Conservancy began to understand enough about the concerns of the watermen and local seafood processors to talk to them about the potential of an enhanced marketing effort, emphasizing the clean waters of the system the Conservancy was trying to conserve. During those same years, TNC was helping residents form a community group that would wrestle with the issues of the broader economic future of the southernmost of the two Shore counties. The Conservancy looked across the country for a facilitator to guide and add to the process and finally recommended the Corporation for Enterprise Development (CFED). With assistance from TNC and CFED, the community group ultimately developed into the Northampton Economic Forum and produced a remarkable set of guidelines for continuing development of the county. The Forum published a report titled "A Blueprint for Economic Growth," which explains that the Forum's work is conceived as a way to follow through on the goals established in the county's Comprehensive Plan. Five major goals were articulated: conserve the county's natural resources, preserve the county's rural character, pursue economic self-sufficiency for all citizens, provide adequate public services for all citizens, and pursue and establish a diversified economic base by supporting agriculture, seafood production, tourism, and industry compatible with the goals and objectives of the county's Comprehensive Plan.

The Virginia Eastern Shore experience coalesced for the Conservancy a broader emerging sense that only the knowledge, vision, and commitment of the people living in and near a natural system are, in the end, going to protect, defend, and save the system. Apropos of that, the Northampton Economic Forum's report explicitly reflects in a summary goal statement the community's intention to preserve "the globally significant natural resources, character, history, and quality of life" that "are Northampton County's special strengths." The community had the solutions and the direction within it. But without the catalytic efforts of a few community leaders who were in turn catalyzed by their relationships with a conservation organization that was itself struggling to develop a new approach, the community might well have turned in a different direction. Absent the community's express intention to preserve the farming community, forces of disintegration would be overtaking the farming life as rapidly as they have in so many areas in coastal Virginia. Without the community's attention to achieving a different result, the poor of the county would be priced out of places to live or be offered residence in soulless places that sever their connection to the land and water traditions of their community. Without community focus on the goal of preserving resources, character, and quality of life, the soil remediation plant that promises a dozen jobs in exchange for a little or a lot of degradation of water quality may be too attractive to resist; leaders of rural communities feel pressure to do *something* to improve the prospects for the young people of the community, for those few who haven't left a place where opportunity seems absent and prospects for improvement poor. Without the attention, too, of local producers and regional marketers, the products of an environmentally sound land and sea cannot be distinguished as such and receive the market premium that consumers will award for their quality and local origin and, perhaps, for the holistic philosophy of production that generated them.

Each successive step forward on the Eastern Shore stretched the boundaries of TNC's capability and that of its partners. As that conservation project has grown and branched, the Conservancy has, with others, created new institutions to take on increasingly diverse and demanding work: affordable housing issues, understanding the ecosystem, and assessing the implications of various proposed developments on it. Among the most interesting of these new organizations is one that will identify, develop, and market products that reflect Eastern Shore traditions that can be produced in an environmentally compatible way. It is called the Virginia Eastern Shore Sustainable Development Corporation (VESSDC), and it has been designed to be the vehicle by which the community reaches a market that will (or can be persuaded to) pay a little more for products that are part of a credible effort to preserve the character and quality of the place.

The VESSDC is a for-profit corporation in which the Conservancy holds a substantial equity interest. The corporate charter specifies exceptional purposes: "to develop and support products, business ventures, and land uses that enhance the local economy, achieve community goals, and preserve the natural resources on the Virginia Eastern Shore. . . . The corporation will assure through its structure, governance, and operations that economic profits are not secured at the expense of environmental or community degradation." The charter also creates three classes of stock, one of which can be made available for equity participation in the business by segments of the community including, ideally, the producers of the products the business will market.

The Conservancy decided to organize a for-profit corporation rather than pursue the same goals through some other form of business organization for a variety of reasons, three of which bear special mention. First, TNC believed that a bottom line of the kind with which people inside and outside the community are familiar and comfortable would allow the effort to be better understood, more like the development that communities traditionally look for. Second, the Conservancy thought the for-profit model would enhance the effort's potential to be (and reinforce its own resolve to create) a replicable model. Third, it believed that the for-profit form, specialized though it was, would interest investors in providing capital to empower TNC's conservation efforts that it could not otherwise attract.

The marketing corporation has indeed brought new capital to the project; that alone suggests that the intuition with regard to the psychology of the form was correct. It will be several years before it is known whether the financial predictions of the pro formas have been met. They were persuasive enough to attract a superbly qualified president for the corporation who will try to make them come true.

Those pro forma financial statements did not, it is important to add, predict the kind of profit that a start-up business usually has to promise to attract investment in the face of the risks that attend all start-up companies. And, of course, TNC added to the risk load with the environmental and social restrictions included in the corporate charter. The latter make it unlikely that soaring profits will ever be obtained. The corporation will market only products that are consistent with the stated ecological goals. It will favor those that reinforce the conservation mission. The effort will be human and local in scale, and it has bound the standard corporate purpose of maximizing profit to requirements that the corporation serve the community's social and environmental requirements and aspirations.

Society needs institutions that will accept the challenge of serving all these masters. They are not as easy to create and capitalize as other businesses. But there is enough money out there to bring the first of them to life. And there is good in developing a class of investors who will accept reasonable returns distinguished by their sustainability from companies that demand a little more of themselves on the journey to the bottom line.

Conservationists who conceive of, create, and capitalize for-profit businesses as a conservation tool have strayed a long way from the terra firma of parks and preserves. Indeed, if TNC had not carefully defined system, observed stresses, and analyzed sources, it might be dangerously far from shore. The waters in which the project is now sailing are uncharted, but the course it is following is far from capricious; the helmspeople are seasoned, and they have a clear idea of where they want to go. The Virginia Eastern Shore project embodies the holistic conservation approach—ecology, community, and economy—introduced in chapter 5.

* * *

When conservation project leaders begin to get deeply involved in community and economic affairs, they must have a clear sense of themselves, their purposes, and their priorities. That clear sense is the foundation for establishing and maintaining healthy relationships with the people who affect and are affected by the lands and waters that are important to the project. Community development rhetoric might lead one to believe that a conservation organization that wants to work with a community ought to begin by simply asking the community what it wants to accomplish—and then proceed to help the community accomplish those things. The Conservancy has been variously criticized and ridiculed for showing up in communities with the stated agenda of biodiversity preservation and the unstated intention to pursue the agenda "whether you like it or not." But acting as if there were no agenda will satisfy neither the community nor the conservationist. The community does not and should not believe us when we say: "We're from the Conservancy and we're here to help you." The Conservancy is a conservation organization. It exists to conserve biodiversity, it's in that community to conserve biodiversity, and an honest relationship between the project team and the community cannot be built on any other premise.

It seems to me that the right approach is to be respectful enough to expect that the community can handle the truth. The conservation organization has to begin by explaining that it has conservation objectives. Often the instinct to find common ground will move the conservationist toward downplaying the real purpose of his or her organization's presence in the community, but clarity about, if not a restrained pride in the mission, is a critical foundation for relationship. The message (in effect and perhaps over several meetings) is, "We're committed to the conservation of the barrier islands, the fishes and mussels in the river, the migratory corridor. You live here. You can tell us a lot about this system, and you have reason to care about it. You're more important to its future than anyone else. We know that you want and need and care about other things too, and some of them seem to be in conflict with our goals. Help us to understand what you want and need. We will work with you to accomplish those of your goals that reinforce our goals, and to the extent we can, we'll work to accomplish the most important of the rest of your goals in ways that are consistent with the conservation objectives we have." When those words are followed by actions that demonstrate their sincerity, the relationship can move forward.

Because the Conservancy has been exceptionally narrowly focused, its project teams (like most conservation project teams) have very little experience in working on "the rest of" a community's goals. It is surely important for the project team and the community to understand that there are limits, that "to the extent we can" means that limitations on resources and the need to set and pursue priorities make it impossible to provide help on every issue of community importance.

In a project established to help protect one of the last sizable remnants of Brazil's extraordinarily diverse Atlantic rain forest, the Conservancy and its local partner had occasion to facilitate the provision of an improved system of health care—that community's highest stated priority. Health care is critical business, but not the Conservancy's business. Even so, the Conservancy put together a project focused on providing health care services, and it has shown very promising early results. The resources that made the project possible were available from a donor that truly was not interested in supporting direct conservation strategies. Project team members believed that attending to health care needs would build their relationship with the community. That rationale is always available, always easy to cite; it can explain a team's decision to expend its resources on any community priority. But it may not justify such a decision. Early evidence from the Atlantic rain forest indicates that talking health care was indeed a way to begin talking conservation. In general, though, there are limits to conservationists' ability to become immersed in communities' problems and goals without exhausting their resources and losing track of their original objectives.

Indeed, in that Brazilian coast project, one of the issues may be how to move gracefully away from the natural perception that, in spite of protestations to the contrary, the Conservancy and its Brazilian partner organizations will be good all-purpose sources of resources for educational infrastructure, programs to alleviate poverty, and a long and diverse list of legitimate community needs. The team needs instead to cultivate the community's understanding that TNC and its partner are in the region to conserve Atlantic rain forest and that they see community and economic issues through an ecological lens.

The conservative course is to decline to get involved with community priorities that cannot be addressed with strategies that alleviate priority ecological stresses. The consequence of that narrow focus may be slower progress with the community; some community members will find us irrelevant, unrealistic, and generally more a nuisance than a help. We can get past that if, having structured our first interactions with the community to be clearly and directly related to our conservation objectives, we establish a history of meeting our commitments in such interactions. That kind of performance and fidelity will mitigate the community's initial disappointment in the narrowness of our focus. We will also be preparing the way, should we decide to work on health or transportation, for the expansion to be perceived more as a welcome exception than as the undertaking of the first of a series of obligations.

It's worth noting, if not entirely unexpected, that resident field managers of projects are inclined to find connections between community priorities and conservation goals that appear tenuous to supervisors in off-site offices. The final say belongs to the on-site manager in a well-managed effort, but a good field manager will also listen to the cautionary voice of the colleague who can observe the work sympathetically but from outside the fray.

* * *

If it is hard to decide when to pursue a strategy that will not directly address identified stresses, it is even harder to decide when to adopt as a strategy a modified form of a major threat. There are two possible reasons for doing such a risky thing. First, there may be an opportunity to garner for conservation purposes some of the financial power associated with the threat. Second, if we can adapt the threat and preemptively employ it in a way that is less damaging to our conservation purposes, we may begin to show for more general application that there is a more desirable way to meet the needs or wants the threat exists to address.

Along the lower Cosumnes River in central California, the Conservancy has, as mentioned earlier, an interest in preserving and restoring riparian forest, grasslands, and wetlands. The stresses include fragmentation and direct destruction of the system, with agricultural and residential development as the source. One of the keystone tracts, 4,300 acres in size, had been optioned several years ago by a developer who applied for approval of a general plan calling for construction of 4,500 homes and several hundred thousand feet of commercial and retail space. California's economy slowed down, however, and ground breaking was delayed to the point that the developer finally let the option lapse.

It was the Conservancy's assessment that absent intervention, development was unlikely over a three-year time horizon—and highly likely within ten years. The 4,300-acre property was available to be purchased, though the price was a slightly breathtaking $12 million. And the larger system TNC wants to conserve is a hundred times the size of the tract in question. TNC and the natural resource agencies that are active in the valley will purchase certain properties within the Cosumnes system. They won't, however, be able to conserve the system by purchasing all of it.

The project team might have waited and watched the 4,300-acre tract and hoped that by persuasion and advocacy it could blunt the effect of the development that would come when the market bounced back. It chose instead to explore the possibilities of paying for part of the desired conservation result with a new tool, a kind of preemptive development designed not to maximize profit but to minimize environmental effects while covering costs. The potential seemed sufficiently powerful that the project team picked up an option on the 4,300-acre property and is considering a development project that would include 50 to 200 units on 700 of the acres; the building plans would preserve much of the natural character of that acreage. Eighteen hundred acres would remain in agricultural use, with some 250 in cultivation and 1,550 in dry pasture that would leave vernal pools undisturbed. The balance of the property would be purchased outright and managed as a traditional natural area.

Because the Conservancy's board, staff, and membership are reluctant to put the organization into the position of developer, even a developer with conservation purposes, and even more importantly because the Conservancy is *not* a developer with the experience of turning good development plans into a good development, the original plan was to sell the option on the land to be developed once the project team obtained county approval of the Conservancy's environmentally responsible build-out plan. The final stages of the development project would be left to the buyer.

The real estate market in the region remains soft, and infrastructure costs for the development are high considering the limited return, so it isn't certain that the Conservancy will take the project to that ready-to-develop state. It remains to be seen whether this kind of restrained development will be of interest to the development community as it is now constituted or whether new business institutions will have to be created to carry such a project through to a successful conclusion.

In any event, as it moves ahead, step by step, the Conservancy is afforded unique opportunities to educate regulatory and elected officials and the community in general about its concern for the system. Equally important is the effort to learn what it takes to create economically viable alternatives to canned development plans that ignore ecology—alternatives that will be better for the community over time. When an economically competitive, ecologically sensitive project is successfully completed, it will challenge community and government perceptions about what is possible with respect to all subsequent development.

*   *   *

The decision to work development into conservation plans wasn't so hard at the Cosumnes, where any fair assessment of the future of land in the Sacramento-Stockton corridor would conclude that development was coming. Certainty is a little harder to come by around the Gray Ranch, three hours from Tucson and three hours from Las Cruces, New Mexico. In the Borderlands region, you can see a new subdivision here and there around Rodeo, New Mexico, or in Douglas, Arizona/Agua Prieta, Mexico, and there is some interest in inexpensive housing development driven by the smelter at Playas, New Mexico. A ranch here and part of a ranch there are cut into ranchettes or lots—the market of buyers is thin but not imaginary. Just to the north and west are the Chiricahuas—beautiful but already divided and subdivided wherever they are not federally owned, lost mostly to second-home development.

Development of the kind that drove the Conservancy in the Cosumnes Valley will be a long time coming to the Borderlands area around the Gray Ranch, if it ever comes. But perhaps the critical and distinguishing biological characteristic of this southwestern high desert system is its minimal fragmentation. Development, even of the desultory kind currently visible around the area, is therefore a serious threat. It also comes accompanied by all kinds of subsidiary threats associated with accommodating more late-twentieth-century humans in the landscape—pets gone feral, exotic plants, the effects of water recovery and treatment. Resuscitation even in the rather unlikely event of eventual abandonment is likely to be slow and incomplete.

The Conservancy has therefore begun to work on a preemptive strategy in the Borderlands area as well. There, instead of turning to low-intensity residential development, it is working to stabilize landownership and land use in the pattern that now exists, a pattern that has emerged from the predominance on the landscape of relatively large ranches that are now managed just about exclusively for raising cattle.

Like limited residential development, that choice brings with it some clear environmental costs. There were probably no hoofed animals of the weight and grazing habits of cattle in that landscape before European American settlement. If we were to stretch a point and compare cattle to bison, we would have to go back to Paleolithic times and a different climate to find them there. Besides, the differences between bison and cattle from an ecological perspective are well documented. But that landscape has hosted a variety of uses, including a community of several thousand Meso-American Indians only five hundred years ago. It looks wild, and it is. It isn't pristine.

The choice, then, is between supporting and promoting a conservation compromise that is believed to be manageable—ecologically conscious grazing—and taking a risk that the unmanageable invasion of subdivision will never arrive. It can be and has been argued that by working with local ranchers the Conservancy is applying its resources to the prolonging of a damaging use that might well go away if left alone. But, if that point opens the argument, the obvious rejoinder is: after grazing, then what? Neither this land nor nearly any other unprotected land will long be off the path of other kinds of development within the means and desires of a growing population with enormous financial resources. North American entrepreneurs abhor a vacuum even more than does nature.

I believe that the Conservancy's decision to find common ground with and support the community of ranchers in the Borderlands, an interesting one from a rural development perspective, was correct even if analyzed as if conservation were a goal that could be pursued in a vacuum. The issues surrounding that kind of decision, however, are real and complex. Whenever a project team selects a conservation strategy that is linked with development, the purported benefits ought to be regarded with skepticism and the risks regarded gravely.

A couple of years ago, in a conservation plan review session, a Conservancy project director outlined plans to meet the coming challenge of development in the new Micronesian nation of Palau by promoting development alternatives compatible with the preservation of that island nation's incredible marine resources. TNC's director of science at that time, having heard the economy of Palau characterized as one of "affluent subsistence," asked why the project team didn't instead try to promote the maintenance of a subsistence way of life in this Pacific Eden. The answer is that such a strategy is unlikely to work any better than it did in the biblical Eden. Development was coming to the islands. Investors saw money to be made around those warm, clear waters; natives saw too many things they wanted in the developed world to maintain a subsistence lifestyle, "affluent" though it seemed. But the science director's question was appropriate. The question always has to be asked.

Even if development is inevitable, you don't want to hurry it, unless you must concede that battle to have a chance at winning the conservation war. Consider the response of the leaders of a remote village on a riverbank in eastern Peru when asked what they wanted most for their village: a satellite dish. With that acquisition, the world and all its goods, its proudly displayed decadence, and its fatally skillful marketing will come to that village. Are they likely to resist the imperial forces of global commerce and decide instead to maintain their subsistence lifestyle, however idyllic and harmonious it may seem to us as visiting observers? Life can be made easier with powerboats and sonar to search for fish and high-powered rifles with which to hunt. Life can be made more interesting with prepared foods and someone else's recipe for liquor. But those things require cash and the desire to have cash changes everything.

That doesn't mean there's no hope of preserving important elements of traditional ways of life and traditional values. The Palauans and the Peruvians retain pride in and a desire to maintain the best of the old ways as well as an interest in the new. That pride must be honored. No one, though, is going to succeed in trying to freeze them in time or block their access to the global village. Moreover, we don't have the standing to try.

It can well be asked whether we're losing or winning when we take an approaching environmental assault and pull it along, albeit on a different—and, if we plan well—substantially less damaging track. The answer is both. If we are lucky and our judgment is good, we'll minimize the loss of the natural world. That isn't exactly winning; it concedes some loss. But it is a win of sorts if the compromise to which we accede preempts a much greater loss.

All strategies for managing development pressure aren't equal. The best strategies reinforce a healthy relationship between the community and the ecosystem. In recent years, for example, it has become evident that there is economic value in nature as nature. That value is usually realized as revenue from nature travel or ecotourism. Said to be the fastest-growing segment of the biggest industry in the world, ecotourism involves taking people to places that still feel wild and often providing some natural history education along the way. The Conservancy is exploring ecotourism at a number of places, including the Cache River in Illinois and Palau. Ecotourism is also prominent in its plans for the Eastern Shore of Virginia, where it is conceived of as a way to enhance the market for limited residential development and locally produced products.

There are places where nature travel or outdoor recreation will produce more income than any likely economic alternative. Even in those places, only a very carefully planned and managed program offers both sufficient revenue to be attractive and prospects of retaining ecological and cultural values.

It barely needs saying that ecotourism won't work everywhere we need to do conservation. First, everything that needs saving is not of broad enough scenic appeal that economically viable numbers of people will travel to see it. A second problem with ecotourism is that accommodating economically viable numbers of people, and the infrastructure required for doing so, will be destructive to systems that are fragile or biologically isolated. A third is that ecotourism tends to generate relatively low paying, service-oriented jobs. The opportunity to work as a waiter or a cleaning person is not, on the whole, going to bring people to their feet in a spontaneous expression of support when plans for the tourism facilities are announced. We aren't, in short, going to build a new and sustainable world economy based on guiding one another around nature preserves.

In the course of reviewing many plans for the conservation of systems, The Nature Conservancy has considered many different kinds of development. I have deliberately avoided the use of the now-ubiquitous "sustainable" label for development of the kind to be considered here. For system conservation, it seems to promise both too much and too little. I'll explain.

Sustainable development, according to the most frequently cited definition, "meets the needs of the present without compromising the ability of future generations to meet their own needs." The definition begs the really important question: "What are needs, and who says so?"

Economist Herman Daly has provided a somewhat more rigorous definition of sustainability. He suggests that an activity may be termed sustainable if it uses renewable resources no faster than their rate of regeneration, uses nonrenewable resources no faster than sustainably renewable substitutes are developed, and pollutes no more than the environment can assimilate.

I know of no development of substantial size that could pass Daly's thoughtful set of tests. It is such a patently sensible way to judge activities that we ought to ask ourselves why not. The answer is that unsustainable habits are deeply ingrained in our way of living. Whenever a process associated with economic activity uses, for example, petroleum, it would seem to fail the Daly test. For some of the strategies we want to use to reduce immi-nent stresses, we may be willing, then, to ask for something less than sustain-ability. It asks too much; what indeed we must achieve in the long term is sometimes more than we can give in the short term.

It can also ask too little. We can imagine, as some conservationists have, a well-managed tree plantation that is sustainable by the cited defini-tion. (Processing and transportation to markets too? Well, we're imagining this example.) If that plantation replaces native vegetation with fast-growing exotic stock, if it replaces complex natural systems with simplified artificial ones, if it takes supportive landscape context and replaces it with fragmenta-tion, we might still judge it sustainable, but it would surely be incompatible with efforts to conserve the natural system it replaced.

I prefer, for project planning purposes, the label and standard of *compatibility* for development, and I suspect that the Conservancy was the first in the conservation movement to make such use of the term. Compati-bility means simply that the development in question can be planned to pro-ceed in a manner that is consistent with conservation of the system. An oil well isn't a sustainable enterprise. Carefully managed, though, it might be compatible with the ecological requirements of the system in which it is located. Sustainability is a critical long-term goal. Compatibility is an imme-diate necessity.

There is a nearly infinite variety of kinds of development that could be conducted so as to be compatible; these can be grouped usefully into three broad categories. The first can be called "unrelated" compatible develop-ment. Few land uses other than grazing will, for example, support people who want to live on Oregon's high desert. But grazing, as currently con-ducted, is one of the problems facing the ecosystem. A Conservancy project team working on conservation of the system speculated about attracting as an economic alternative clean, "footloose" business to the area, utilizing the area's outdoor values as the magnet. (Footloose businesses are those that produce products or deliver services that don't have strong locational con-straints, such as computer software or polling and market research services.) The Wilderness Society has done serious work investigating that idea in the greater Yellowstone region.

While the idea is certainly worth considering, it will not be a strategy that works as often as we would like. There are not that many viable busi-nesses that are amenable to moving, and everyone everywhere wants to attract those that exist. (The statistics for smaller businesses may be more promising, but Greg Low, a brilliant land conservationist, a Conservancy vice president, and coleader of the Virginia Eastern Shore project, cites figures showing that with respect to the traditional target of economic development efforts, major industrial plants, there are 500 relocations each year, most of these not the clean businesses that The Wilderness Society envisioned, chased by some 30,000 communities.) Also, even assuming we could direct the development that follows new business to the places we want it and away from the places we don't want it, there is still a gap. For footloose businesses, the environment is the magnet—the open space, the outdoor and recreational values. The business, by definition, could be conducted anywhere and does not itself depend on the health and integrity of the natural resources of the area. The area has to look nice, and recreation has to be accessible. That's something. Of course, every place we want conservation success doesn't look nice enough to attract footloose businesses. Even more important, an attachment to appearance that isn't more fundamentally grounded does little to stabilize land use and could in fact destabilize it with housing needs for the footloose staff people that follow the footloose business. Without an economic relationship that encompasses more than scenery, it is hard to see how chasing clean businesses will build in the community the land ethic we want for ultimate success.

Perhaps the employees of these companies will become deeply attached to their new homes; perhaps they will care broadly enough about the outdoor values to become a political force for preserving ecological integrity. Or perhaps they will pack up for the next fashionable location when the mountain bike rides start to be less pleasant and the fishing less productive. The denomination of the businesses we are seeking itself evokes an element of a generalized environmental problem. Too many of us don't belong anywhere; we don't put down roots and commit ourselves to the well-being of the place we are. We're footloose. Without roots in the land, the deeper, cultural understanding of ecological processes that underlies a knowledgeable social commitment to conservation can seldom develop.

A somewhat more interesting development strategy might be called "related" compatible development. It generally involves making adjustments, sometimes minor, sometimes major, in a land-related economic activity that typified a system when we entered the scene as conservationists. The restricted residential development the Conservancy is pursuing on the Eastern Shore is related compatible development. TNC is trying in that place to show that well-planned, low-density development can compete, all benefits considered, with the high-density residential development that land was attracting.

If we are successful in finding farmers who want to plant low-input, low-erosion alfalfa instead of annual row crops along the Big Darby in Ohio, if we develop with our partners a model cattle-growing regime in the Borderlands, if pioneering efforts take hold and it becomes standard practice to reflood California rice fields instead of burning the stubble after harvest so that they provide habitat for waterfowl and shorebirds as well as off-stream water storage, we will have done some related compatible development.

An even more attractive kind of economic activity could be called "allied" compatible development. A relatively small-scale example is the grazing of bison being conducted at the Conservancy's Tallgrass Prairie Preserve in northeastern Oklahoma. It's small-scale because it has not been planned as a truly viable business venture—the revenue would not, for example, service capital costs. But TNC will generate substantial income through an activity that benefits the ecosystem it is conserving. An effort of like kind is being planned at the Conservancy's Niobrara Preserve. In that Nebraska Sandhills refuge, the preserve staff not only manages bison but also hopes to address an overabundance of cedar by transforming some of it into the raw material for a fence post business.

The Conservancy discussed an even more interesting effort on Virginia's Eastern Shore. The idea was to get members of the fishing community, locally called watermen, into a cooperative venture that would take shellfish from the biosphere reserve waters, perhaps at a reduced rate, while making up the likely difference in volume by commanding a premium price for "bioreserve oysters"—taken from exceptionally clean waters in a biologically and environmentally sound manner. If TNC can help successfully launch that venture (it's on hold, waiting for, among other things, the oyster population of the mid-Atlantic to recover from a disastrous virus and other assaults), the watermen will be strong allies for keeping those waters clean and will have strong incentives for regulating their take very closely. In such a business, jobs and economic success are directly linked to the health of the ecological system that is the object of conservation attention.

It is tempting to assert that this kind of business creates links between the economy and the environment. It does not create links; those links exist in every human endeavor. Allied compatible development makes them so clear and so immediate that their fundamental power becomes evident.

Conservationists have a great deal to learn in this area, but it is a realm of great promise. One universal difficulty with compatible or truly sustainable development is that it's almost axiomatic that short-term returns will be lower than they would be with a profit-maximizing approach. That presumably makes it less likely that people will choose to work or invest in such ventures. But there are circumstances under which they will make such a choice. The Conservancy's effort to set up a sustainable timber operation near Yellowstone National Park in Montana, for example, demonstrated that at least when faced with the choice in its starkest terms, communities may well choose sustainable alternatives in spite of the short-term economic trade-off.

That project began with an attempt to establish a new conservation area by purchasing some twenty thousand acres of high natural quality in the Yellowstone region from Plum Creek Corporation. As a successor to the old Union Pacific, Plum Creek owned hundreds of thousands of acres in a checkerboard pattern across the West, originally transferred by the U.S.

government as part of a package to induce Union Pacific to build a railroad line to the West Coast. The acreage TNC wanted was part of a Plum Creek management unit that included some 175,000 acres east of Ennis and north of Belgrade, Montana. That portion of the forest which was accessible to logging had supplied a stud mill in Belgrade.

Plum Creek not only wouldn't consider splitting off the priority conservation land, it insisted that the mill had to be part of the deal, too. After a little background work, the project team learned that the life expectancy of the mill was less than three years because the timber supply available to economically feed the mill was so far diminished. Plum Creek would have been very happy to have someone else close the mill; perhaps it was particularly fascinating to contemplate the notion that it would be TNC.

The team members thought about the land, the conservation need, the timber supply, and the possibilities. They engaged consultants to evaluate the costs and benefits of a different approach to utilizing some of the trees in the region and returned to the bargaining table prepared to discuss the purchase of the entire tract and the mill. The highest-quality parts of the land package would be put under public or private conservation protection, while most other lands would be sold without restriction to the Forest Service or traded as outdoor recreation land to Montana Fish and Game. A small amount of land near the Big Sky ski resort would be sold for development; its particularly high value could finance the conservation of a lot of other land in the package. The land and timber that wasn't designated for strict conservation, transfer to public agencies, or sale would support a sustainable timber operation that would be established and operated with the local unit of a medium-sized timber company as a partner.

Sustainable levels of timber harvest meant a reduced flow to the mill, the effect of which would be mitigated by investing about $1 million in new equipment that would support production of more intricately milled specialty products—making it possible for the plant to add value substantially in excess of that associated with the production of two-by-fours. The pro formas made it clear that some jobs would be lost; the refitted mill was expected to be able to retain about 60 percent of its workforce.

The loggers and millworkers, the broader community, and government officials knew that absent some changes in approach, the mill's future was limited—it would close in one to three years. Community leaders endorsed the plan TNC put forward. In those circumstances and to that community, sixty sustainable jobs were better than a hundred unsustainable.

In the end, the Conservancy was outbid for the property by a half a million dollars in a deal of more than $20 million. The plan was never put into action. Even so, the questions answered at Belgrade weren't abstract or hypothetical. There was a real choice to be made, and the way that community chose provides a lot to think about, as well as a little encouragement for those worried about the future of the planet.

As a footnote, the Conservancy should have bid more (and I was one of those who decided we'd reached our limit). Timber prices went up. There isn't a happy ending to every story. (Actually, the end to this one hasn't been written. The Conservancy and others are still working with the new owner in an effort to get the priority tracts protected.)

*   *   *

The beginning of this chapter emphasized that the objective of strategy is to address the sources of priority stresses, and the examples given have included some pretty bold plans. Most of the strategic plans the Conservancy has reviewed set forth less ambitious strategies. Team members haven't felt confident enough to commit themselves to pursuing the ideas that might change the dynamics of the community.

There is nothing either surprising or necessarily wrong about that. If there is time, or if the conservation project staff and resources are such that the program is as likely to strike out as to hit a home run with the bigger swing, it's better to plan to make contact, achieve some smaller successes, and prepare deliberately to launch strategies that promise more.

To any reader who knows the Conservancy or has paid attention to the previous text, it won't come as a surprise that TNC people think perhaps the best strategy for small success is acquisition of the most important lands within the system. There are both obvious and subtle reasons that is so. Obviously, acquisition of land confers the firmest control, preserves the most options, and displays the most tangible results possible from conservation action and expenditure. Less evident is that real property ownership is an honored and protected status in the United States and in many other cultures. With ownership of property comes a sort of moral right, like it or not, to comment on civic questions and participate in civic debate as one who belongs. It is a different status in kind and quality than interested bystander, do-gooder from the capital, or outside agitator. Property ownership doesn't bring acceptance in the community; it can generate suspicion, fear, and anger. But for all that, there's a widespread willingness to grant to any landowner the right to protect and defend one's own, one's investment.

As noted earlier, much of the Conservancy's energy since its incorporation in 1951 has been devoted to land acquisition, and the result is a proud legacy. The need for that work hasn't diminished; indeed, as more and more natural land is converted to intensive human use, and as open space dominated by agriculture or silviculture is fragmented and converted to suburban and second-home development, there's a greater need for land acquisition for parks and preserves than ever before. We need more state, federal, and local acquisition, and we need more land trust and Conservancy acquisition.

The underlying premise of this book, though, is that the question of how much biodiversity, how much integrity of natural function, we leave in the world for our children and grandchildren will not be decided with land acquisition. In the 1950s, the Conservancy advertised that it was in the business of preserving "living museums of primeval America." Unless we can find strategies that affect the use of that part of the landscape—much the greater part—which isn't acquired for parks and preserves, that now quaint-sounding purpose will be a precise statement of what it is that we, the Conservancy and society, have done. We need to do more than that.

There's a good deal of middle ground between capitalizing for-profit sustainable development corporations and acquiring land. The most frequent makeweight middle strategy that has appeared in Conservancy plans is education.

Many of us have been heard to say at some time or other that education is the ultimate answer, by which we usually mean that if others just knew what we know, they'd behave themselves, and our systems would be saved. Education as the term is often conceived in conservation is a facile strategy that borders on arrogance. We have devoted an awful lot of good effort to producing a population that calls itself environmentalist. We have succeeded by some measure in educating the public without halting our collective march toward destruction of natural systems. Not that there aren't bright spots. Indeed, we have passed the Clean Air and Clean Water Acts, the National Environmental Policy Act, the Endangered Species Act, and the Wilderness Act. That legislation has certainly produced positive results for the environment, but it was law long before the environment pulled the 80 percent–plus numbers we see in polls today, long before environmental education was a common part of grade school curricula. And the widespread environmental consciousness, such as it is, that we have now, what does it mean, what values are widely shared? The congressional assault on the foundation of U.S. environmental law in the spring of 1995 didn't produce notable public outcry.

Environmental consciousness as expressed in the existence of organizations devoted to the cause dates back about a hundred years in the United States. The recent membership growth of environmental organizations in the nation has made more of them bigger than ever before, but it hasn't been a uniformly steady road up. The Izaak Walton League, now comparatively small, had more than 100,000 members in the 1920s, when the Audubon Society was supported by fewer than 10,000 and The Nature Conservancy was a committee of the Ecological Society of America.

Audubon reached the 100,000 mark in 1970, and its magazine commented in that year of the first Earth Day, as historian Stephen Fox has noted, that "now, suddenly, everybody is a conservationist." Recent polls in this country consistently show that 70 to 80 percent of respondents not only consider themselves environmentalists in the broad sense but also say that they specifically support conservation positions on a wide variety of issues. An international Gallup poll taken near the time of the United Nations Conference on Environment and Development at Rio de Janeiro in 1992 suggests that sympathy is shared worldwide, in developing as well as developed countries.

But what does that support mean? One gets the feeling that a slight rewording of the questions might elicit a very different set of responses. Our behavior as a national and world group doesn't seem to suggest that our asserted consciousness runs very deep. Maybe we have just taken the first steps on a longer road toward new environmental values. It is, I suppose, commonplace sociology that changed language precedes changed values, and the existence of a value is demonstrated not by language but by action.

At this point, there is still a big chasm between our language and our actions when it comes to preserving the integrity of the ecological systems of the planet. I think new, additional, or continuing efforts at broad-based consciousness raising have an enormous challenge before them if they hope to change that. And while I wonder just exactly what past efforts have accomplished, I believe we have no choice but to accept the challenge. Among the most compelling reasons for doing so is the potential for changing the way people think about what they buy—perhaps the least successfully explored aspect of current environmental consciousness. There is potential to actualize a market that demands environmental sensitivity if we can figure out how to release that potential. Environmental education for that purpose and others is probably a little like advertising—at least 50 percent is wasted, but you just can't tell which half.

Because environmental education is so important, it would seem to follow that it would be a fine idea to include strategies for educating "the public" as a prominent feature in plans for protecting natural systems. It may indeed be that there is an identifiable group with known control over a stress-causing source that can, with a well-defined education program, be induced to create less ecological stress. That sort of education program, however, would not be described as educating the public. There is a role, in an ecosystem approach to conservation, for education that may consist of a series of face-to-face meetings with landowners, civic leaders, businesspeople, and people who make their living outdoors. It may be useful to take some of those people into the field. It may be a good use of resources to sponsor some field days for the children of the community or to develop project-influenced curricula for local teachers. But system conservationists should adopt only the lowest-cost tactics aimed at trying to influence the "public"—that is, everyone. That task has to be left to the increasingly large corps of educators who make it their principal business.

An organization known as RARE has sponsored educational programs that have achieved notable success in the Caribbean. Those educational campaigns have sought to stir national pride in the environment by developing campaigns around organisms that are well known and are bound up in some long-standing way with national consciousness—parrots and butterflies on the islands, the toucan in Belize. The effort seems most often to be directed at cultivating general receptivity to a wide variety of environment-protecting actions. It is frequently cited as a successful approach, and it has its analog in this country in the way we have used the image of the bald eagle. But there is a difference between such efforts and system conservation. If the audience for these efforts has not been selected with the discrimination advocated here, the message has been. It's simple and visual, and it presents a familiar and positive image in a light that is only slightly different. And the result sought is, broadly, general goodwill for a healthy environment. While important, that is not as difficult a task as inducing a specific change in the behavior of a stress-producing source will usually be.

Before adopting education as a principal strategy in a system conservation project, the project team must state clearly the audience and message of the educational effort and the change in behavior—the reduction in a system stress—it is supposed to produce. Education without those constraints demands too much time, money, and effort for it to be an appropriate system conservation strategy.

*   *   *

A different "middle ground" strategy will serve to close the chapter. The concern of the project team at Fish Creek in Indiana is primarily the big-stream fishes and mussels that have found refuge in a relatively small Great Lakes tributary. The principal stress is caused by silt, and the principal source is standard tillage associated with agricultural activities. There was a good deal of interest in conservation tillage in the watershed, but not as many farmers as the system needed had made the transition.

The project manager spent a good deal of time with farm managers and became persuaded that the investment in equipment required was a hurdle just high enough to keep farmers from switching to no-till. When the investment cost was added to the quite human reluctance to take on the risk and uncertainty that comes attached to all change, a broad-based transition to no-till looked, absent a catalyst, unlikely in the short term.

The Conservancy finally decided to try to be a catalyst by using its limited resources to lower the hurdle for the farmers of the watershed. It offered to pay up to 15 percent of the cost of drills and planters, and specified amounts on other equipment, in exchange for a farmer's agreement to create a conservation plan with the Soil Conservation Service; keep cost, input, and yield records; and use no-till technology on at least 250 acres for at least three years. Three thousand streamside or highly erodible acres have gone into no-till during the first couple of years the program has been in operation. As much as 13,000 tons of soil each year will stay on the land instead of riding down Fish Creek.

The key to devising strategies that work is the way the task is approached, and approaching system stresses as problems to be solved has been a successful approach for The Nature Conservancy. As in many problems, there are both variables and constants in system conservation. Houses are detrimental to the natural environment, and so is agriculture, but unless the project team and the community in which it is working are able to exclude or eliminate them from the system, it isn't useful to dwell on the notion that the system would be better off, more natural, without them. In a world that simply will be made to serve the fundamental needs of people, they *are* a constant, and the most useful frame of mind for producing good conservation strategy is to think of them as such. But if their existence is a constant, their impact is a variable. The challenge, then, is to manipulate the factors that surround those constants so that housing, food production, and other fundamental human needs have the least impact on the natural system.

We have in the past mismanaged the environment as we sought to meet our needs, and as population grows, the lessons we once learned about how to avoid the worst mismanagement cease to be relevant and we mismanage anew. In 1995, a spokesman for the South Florida Water Management District said: "The only reason we have been able to live here is because of how we have been able to manipulate nature." His views were newsworthy because nature was resisting manipulation just then—enough people had pushed far enough into the south Florida marsh that moderately heavy spring rains had made life difficult for them. We need to learn that environmental heroism is not finding ways to more effectively manipulate nature but finding ways to fulfill our fundamental needs for food and shelter, as well as other needs and desires, while destroying as little as possible of the natural systems on which we ultimately depend.

Know the natural system, understand the source of the stress affecting it, approach empathetically the community involved and affected by management decisions in respect to it. Accept the boundaries that emerge from fundamental need. Accept no boundaries on ways to meet those needs. Keep working for the winning response to the killer threat instead of the holding pattern or the ameliorative action. Be flexible enough to adjust your approach as you apply it. And hope for a little luck.

What are our chances? A few salmon still make it up some streams against odds that look pretty long. Of course, a lot more salmon start the journey than finish it. We have larger brains but typically less determination than salmon. We sure have a lot of people out in the sea; we have to find ways to help a lot of them start making their way up that stream.

CHAPTER 11

Forests and Trees: Measuring Progress, Defining Success

For years, The Nature Conservancy had the clearest reports of accomplishments in conservation. It reported how many acres it had protected. What could be clearer? Well, that depends on what you are trying to accomplish. The Conservancy professed to be in the business of preserving biodiversity. It wasn't an accident that Conservancy staff and members often said the organization was in the business of conserving land. It measured progress, after all, in acres.

The use of that unit of measurement implicitly suggests that an acre of land protected means something with respect to biodiversity protection. Sometimes it means a lot; sometimes it doesn't mean so very much—even if the acre is the right acre. The Conservancy reports the acquisition of lots of interests in land. But the contribution of that land to preservation of biodiversity depends not only on what biodiversity was found on the land at the outset of its status as "protected" but also on how that land is managed and the status of the ecological processes beyond the boundaries of the holding that are essential to the maintenance of the diversity of life on that land.

The defense of the use of acres "conserved" as a primary measurement is, to make the point in shorthand, that the principal cause of biodiversity loss is habitat destruction, and buying acres for parks and preserves averts the threat of habitat destruction. The point is persuasive, at least as regards the immediately destructive effect of humankind's choosing to convert land to subdivisions, farms, or tree plantations. But the relative absence of the most intensive forms of development doesn't make land into habitat that matters. If we are really trying to measure success, we have to measure more than acres purchased.

Acres are never bad for biodiversity. The purchase of casually chosen land may be open to criticism as a poor choice for allocating scarce resources, but land does offer a certain security and tangibility that makes buying it an attractive tactic to pursue and to measure. But by measuring acres, TNC was measuring activity. That can be useful, and we all need both short- and long-term points at which to check our work. If the activities we decide to measure are well chosen, we can learn something about our progress from the measurements we make. We need, however, to define success more rigorously than that. Success isn't activity, and it isn't even progress. It is impact.

If we are in the business of conserving ducks, we ought, for example, to measure duck populations over five-, ten-, twenty-, and forty-year periods as the most unequivocal part of an assessment of the efficacy of our approach. If duck populations are up, we have prima facie evidence that the conservation effort has had an impact. If population levels have fallen, but demonstrably less than they would have absent the conservation work, there may have been impact of a sort too, but not the impact we want. We're still losing, and we ought to be thinking about a change in strategy. System conservationists should measure the health of the system they set out to conserve and evaluate the strategies they have chosen according to the results of those measurements.

Developing clear measurements that speak to the health of the system is more difficult than might be imagined, and to members of a project team it may not even seem very important. Working in the midst of a project, we develop an understanding of what our goals are, and that understanding tends to become so familiar that it hardly needs expression. That familiar understanding will feel like an adequate definition of success even if it isn't reduced to writing and recorded. But pursuing conservation using an ecosystem approach implies a long-term commitment, a journey of some length in which it is easy to lose the way. The vision of success, that "familiar understanding" of what the project is about, frequently changes over the course of a project. The change will often be gradual and the transitions in project vision nearly imperceptible. The new vision that eventually takes shape may be fully appropriate, but the transformation ought to be conscious. It is certain to be brought to consciousness if the original familiar understanding—initial expectations about what the project was to achieve—have been clearly recorded and are available for reference.

The most desirable measurements of impact are the most direct, and in a system that is relatively simple—a species system, for example—direct measurement is often possible: is the population of the species stable or increasing, or is it decreasing at a slower rate than it was?

When the system to be conserved is a natural community, we need to look carefully into its life history. We can measure progress directly by measuring species composition against what we would expect or desire to see in that community. But many communities have lengthy and complex life cycles. If we find that we are measuring a succession from lodgepole pine to hemlock and larch in a mountain forest community, for example, we had better understand the system well enough to be able to see it as one of five normal states in a lodgepole pine–hemlock–larch community with fire-regulated cycles ranging from 20 to 250 years instead of an unnatural invasion to be vigorously suppressed.

Because of such complexities, we're often driven to seek alternatives to direct biological measurement. Sometimes we can get reliable information about our progress by tracking indirect biological measurements. If we can identify an indicator species—one whose presence confirms the existence of the system we seek to conserve—its status will speak to our progress, at least as long as we aren't managing directly for it, thereby destroying its value as an indicator. Monitoring keystone species, where they exist, can also serve as a manageable way to indirectly measure progress in a conservation effort that aims to conserve a natural community organized around such species.

Also useful, if somewhat less reliable than these ways to measure success, are efforts to track progress by measuring nonbiological phenomena. If we are trying to conserve mussels that we believe are being stressed by silt, the ultimate measurement is whether the mussel population is increasing or declining. We may need shorter-term measurements than the relatively poorly understood and difficult-to-census mussel population, though. One such gauge could be the silt load of the watercourse. Measuring that will tell us the immediate impact of our activity—that is, whether we have reduced the amount of silt. A reduction is cause for celebration, but the conservation impact of our activity and the accuracy of our assessment of stress are ultimately revealed only in the response of mussel populations.

To carry forward the example, because we may not be certain of the source of the silt, we may also want to measure the activities we undertake to reduce silt—how many miles of creek frontage are bordered by filter strips, natural vegetation, or seldom-cultivated crops like alfalfa? How many creek-bordering farms are using no-till methods? How many farmers have been contacted? These measurements say nothing in themselves about the impact we are having—about whether we are saving mussels—but they certainly represent likely mileposts along a road to success.

Perhaps the easiest thing a project team can measure is project capacity—that is, how well we are positioned to do what we believe we need to do. We don't have to invent ways to measure staff, volunteers, support, or money, the common elements of every charitable venture. Measurements of capacity are so familiar that we have a tendency to allow them to dominate progress reporting; they become ends unto themselves. Clear reporting will present familiar measurements of fund-raising, budget, and staff for what they are: capacity for conservation, not conservation.

All of the measurements mentioned have some utility if we remember that our objective is neither dollars raised nor farmers contacted, no-till acres, streamside vegetation, or even silt load in a stream. It's mussels conserved, and all other measurements must be seen as indicators of progress in means, not ends.

Project teams are occasionally reluctant to set up measurements of progress or definitions of success that are end or impact driven because they believe their ability to meet their goals to have an impact, may be affected by factors beyond their control. We need to be wary about acquiescing in that objection; indeed, the conservation community as a whole needs to be more willing to be held accountable for its impact on mission goals in spite of many forces it does not and cannot control. If there really are things affecting the status of the system that we cannot and should not be trying either to influence or to plan around, then the appropriate response is not to shrink from measuring impact and be satisfied with measuring activity but to include a discussion of the factors that are supposed to be uncontrollable in a report of project impact.

Projects that aim to conserve important stopover points in bird migration routes are, for example, frequently structured so as to be subject to effects that might be termed uncontrollable. The surest gauge of conservation impact in such a project is the number of birds that use the site. A project of modest means could quite defensibly decline to take responsibility for the larger migration system—the breeding grounds, migratory corridor, and wintering habitats of the priority species that use the stopover site. If habitat is being lost in any part of the system, the population at the stopover may decline, even if the stopover conservation project is doing excellent work on its part of the system. That, however, does not justify measuring progress solely by a method that considers only work within the scope of the stopover project, such as percentage of habitat on site preserved. Exquisite attention to that battle is no substitute for perspective on the war. The stopover project team must, therefore, gather and interpret information about breeding and wintering grounds and the status of the migratory corridor. More generally, a conservation project team that has consciously defined its boundaries to encompass only part of a system must at the least monitor the status of the entire system, consider the results it achieves in light of events affecting other parts of the system, and periodically reassess its initial conclusion that a limited scope of work represents a reasonably good investment of resources.

As the system to be conserved gets bigger and more complex, the challenge of measuring progress becomes even more difficult because there will be more things to be measured. The cooperation of relevant human communities is often so important, and the economy so primary an influence on their inclination to cooperate, that strategies for conservation of large and complex systems are often social and economic in character and emphasis. Conservationists in such systems ought therefore to be taking the measure of society and economy right along with measurements of ecology.

The rationale for measuring social health and economic performance is similar to that for measuring other things that don't specifically address conservation impact. We must often take some indirect biological and nonbiological measurements to obtain short-term information about the probable effect conservation efforts are having on target systems. Social and economic measurements are similarly indirect and are even more remote from the biological system, but an analysis of stress, source, and strategy that is developed with clarity will reveal things to measure in the community and economy that will tell us of our conservation progress.

Besides, the social and economic goals we decide to measure against will, in contrast to some of our biological goals, hold significant inherent interest to civic and social agencies that would otherwise be unlikely allies for our conservation efforts. Further, as a society we commonly take and keep many measurements of our communities and our economy but far fewer of our environment. If the community leaders to whom we present reports become accustomed to evaluating progress the way we do—putting ecological measurements right up there with the community and economy measurements with which they are familiar through years of habit, that alone will be a noteworthy accomplishment. A community that wishes to bill itself as sustainable (and there are, of late, some that do) *must*, of course, monitor some ecological barometers.

While there are many agencies that gather and evaluate social and economic statistics, the use of such statistics in conservation projects is a relatively new idea. A checklist of measurements likely to be interesting to system conservationists would include population and demographics, educational attainment, housing, indicators of social stress, category and amount of employment, diversity of economic base, and amount of strictly local economic interaction. Specific statistics to track under, for example, the general heading of population might include rate and trend of population increase, age structure of population and trend, social and ethnic composition, and economic status by comparison with a base population (state, region, nation, or world). It would be beneficial to gather and consider a full set of such statistics at the outset of a project and to do so again at intervals of three to five years. From this broad set of statistics the project team may be able to discern trends, receive early warning of coming problems, and gain overall perspective into the workings of relevant communities. The team must also stay focused, though, on the particular issues that originally made it apparent that the conservation project must attend to the human community. That means monitoring more frequently a subset of measures relevant to known community issues that have implications for the conservation work. Lack of economic opportunity for its young people may, for example, be a problem for a community that exists near a natural system we want to conserve. Considering that, we may conclude as conservationists that absent alternatives, the community will be inclined to try to address that economic problem in ways that have negative consequences for the natural system. In that situation, we might have developed a strategy that we hope will address the problem, and we would closely watch age trends in the working population. If we are able to present statistics showing that the average age of the working population has declined while the conservation project has been going on, the community, which will itself be generating a common wisdom about the costs and benefits of incorporating a conservation ethic into its decision-making processes, must conclude at least that it has been able to address its goal of creating opportunities for young people while simultaneously working to conserve its ecological systems; we might even be able to show a more direct correlation between that positive trend and conservation.

The project team will be monitoring a wide variety of social, economic, and ecological statistics at a level of some detail. There is a need, though, for a summary assessment that can be taken in at a glance by associates and supporters of the project team. One way to meet that need, as well as the team's own need for an expressive overall assessment, is to determine the comparative importance of the social, economic, and ecological measurements being tracked and create a weighted scale in which a score of, say, 100 would be achieved with an ideal performance on each of the proportionately weighted areas of interest (see table 11-1). The achievements that would merit a perfect score would be set out in detail at the outset, and a summary description of those achievements would accompany each report. Scoring and reporting of results would probably merit an annual effort. For reporting purposes, there would be an annual score that could be compared with the previous year's score, and for closer analysis, the data that underlie the index would be available for review area by area and measurement by measurement.

Most interested observers would pay attention, I think, to a four-number report that displayed separate community, economic, and ecological indices and a combined project index. A report only slightly more complex than that has been produced in recent years by a group of volunteers in the city of Seattle. The summary version of their report assigns one of three ratings (toward sustainability, away from it, or neutral) to each of forty indicators grouped in four categories: environment, population and resources, economy, and culture and society. Their report gets toward what we want: progress reports that have some of the desirable tangibility of that great old Conservancy acres-protected report but that say something a little more meaningful about what we are really trying to do.

Table 11-1 Detail from annual report of year six results,
Blackacre Ecosystem Conservation Project

Category	Goal	1996 Score	1995 Score
Ecological Measurements	**40 points**	**34**	**32**
Forest cover	10 points	7	8
Average silt load—river	20 points	17	15
Population of rare mallow	5 points	5	4
Presence of blue hawk	5 points	5	5
Economic Measurements	**35 points**	**15**	**12**
Dollar value—value-added production	15 points	7	6
Progress toward import substitution	10 points	2	2
Employment	10 points	6	4
Social Measurements	**25 points**	**17**	**16**
Population	10 points	9	9
Poverty level	5 points	2	1
Economic disparity	5 points	2	2
Affordable housing availability	5 points	4	4

Percentage of Goal Met: Ecological, **85%**; Economic, **43%**; Social, **68%**.

Annual Score: **1996, 66;** 1995, 60.

Backup for 1996 Scores, Blackacre Ecosystem Conservation Project

Ecological Measurements	**1996**	**1995**
Forest cover: 50% cover = 10 points	7 points: 42% cover	8: 45%
Average silt load—river: χ ppm = 20 points	17 points: χ + 750 ppm	15: χ + 1000
Population of rare mallow: 1000 individuals = 5 points	5 points: 1200	4: 900
Presence of blue hawk: Present = 5 points	5 points: present	5: present
Economic Measurements	**1996**	**1995**
Dollar value—value-added production: $10 million = 15 points	7 points: $3 million	6: $2 million

Progress toward import
    substitution: 50% overall reduction
    from base; 5 tracked items = 10 points
	2 points: 22%	2: 20%

Employment: 95% = 10 points	6 points: 90%	4: 87%
Social Measurements	**1996**	**1995**
Population: Stable = 10 points	9 points: +1%	9: –1%
Poverty level: 10% (regional average) = 5 points	2 points: 15%	1: 17%

Economic disparity: Increase by
    5% over base the total earnings
    of the lowest-earning 30% of
    project area population = 5 points
	2 points: +2%	2: + 1.5%

Affordable housing availability:
    1,000 homes meeting code
    at under $75,000 = 5 points
	4 points: 750	4: 700

Table 11-2   Balance sheet: Ecological, economic, and social capital
of a rural coastal community

Ecological Systems	Priority Rank	Weight	Condition	Score (1–10)	Weighted Score
A. Natural communities and species					
1. Estuarine system	Very high	8	B–	6	0.7
2. Barrier island nesting birds	Very high	8	B+	8	0.9
3. Barrier island communities	Very high	8	A	10	1.1
4. Shorebirds	Very high	8	A	10	1.1
5. Neotropical migratory birds	Very high	8	B	7	0.8
6. Waterfowl	Medium	2	B	7	0.2
7. Terrestrial communities	High	4	C+	5	0.3
B. Renewable natural resources					
1. Forests	Medium	2	C	4	0.1
2. Agricultural soils	Very high	8	B	7	0.8
3. Groundwater	Very high	8	A	9	1.0
4. Fisheries	Very high	8	C–	3	0.3
Total		72			
Weighted average			B		7.3

Economic Systems	Priority Rank	Weight	Condition	Score (1–10)	Weighted Score
A. Export of goods and services					
1. Nonrenewable resources	n/a				
2. Renewable resources					
Agricultural products	Very high	8	C	4	0.6
Forest products	Low	1	C	4	0.1
Fisheries (incl. aquaculture)	Very high	8	B	7	1.0
3. Services					
Tourism	High	4	C+	5	0.4
4. Value-added production					
Food products	Very high	8	D	1	0.1
Local arts and crafts	High	4	C	4	0.3
Sustainable technology	Very high	8	C	4	0.6
B. Local trade					
1. Import substitution	Very high	8	C	4	0.6
2. Local goods, value added	Medium	2	C	4	0.1
3. Attraction of outside funds	High	4	B	7	0.5
Total		55			
Weighted average			C		4.3

(*continues*)

Table 11-2   *Continued*

Social Systems	Priority Rank	Weight	Condition	Score (1–10)	Weighted Score
A. Personal and institutional leadership					
1. Citizen leadership	Very high	8	A	10	1.7
2. Civic institutions	Very high	8	A	10	1.5
3. Local government	Very high	8	B–	6	0.9
B. Rural character	High	4	B	7	0.5
C. Local history and culture	High	4	A	10	0.7
D. Public services					
1. Public safety	High	4	C+	5	0.4
2. Schools	Very high	8	B	7	1.0
E. Housing	Very high	8	C–	3	0.4
F. Physical infrastructure	High	4	B–	6	0.4
G. Job skills training	High	4	D	1	0.1
H. Development finance	High	4	B–	6	0.4
		—			
Total		48			
Weighted average			B		8.0

Note: The scoring system is as follows:

Priority ranks (for weighted average)		Conditions (10-point scale)			
Very high	8	**A**	**10**	C	4
High	4	A–	9	C–	3
Medium	2	B+	8	D+	2
Low	1	**B**	**7**	D	1
		B–	6	F	0
		C+	5		

The Nature Conservancy's Greg Low has been exploring the potential utility of adapting a familiar business report, the balance sheet, to five-S conservation purposes. The capital and assets tracked on the conservation project's balance sheet would be human and ecological; progress in reducing system stress and in restoring system and social community health would be shown by comparing successive years' balance sheets. A sample balance sheet for a hypothetical coastal system and the associated rural community is presented in table 11-2. To produce the sample, a variety of natural and social features were selected and assigned qualitative priority ranks and corresponding numeric weights in a geometric progression; their annual condition is expressed both in report card fashion and in a corresponding numeric score that becomes a factor in the production of individual and combined weighted scores. Low plans to test the balance sheet idea, along with a prototype "income" statement, in several conservation projects.

*   *   *

I have included these ideas for measuring progress in large-system conservation projects as much to stimulate readers to develop better-designed, more expressive methods as to share the existing state of the art. It is important that we keep working at developing better measurements.

It is also important to step back after applying any custom-made measurement or set of measurements in a conservation project and affirm that the measurements aren't so sophisticated that they obscure our view of progress on fundamental purposes. Indeed, we need to make a deliberate effort to be certain that basic goals are kept in the forefront of project thinking.

At the end of each planning cycle, therefore, project teams ought to pause and conduct one more exercise, this one designed to ensure that after the choices and compromises that have been made to preserve as many trees as possible, the long-term quest to preserve forest has not been made impossible. We have to step back and recall or reformulate the broad conservation vision that represents the desired long-term state for the system to be conserved. This kind of review can benefit especially from the perspective of a knowledgeable reviewer who isn't a part of the project team.

The five-S planning discipline calls for focus in defining the system, with emphasis on deciding clearly what the conservation object is and on gaining an ecological understanding of it. The discipline requires detachment in identifying the ecological stresses that threaten the integrity of the system and objectivity in tracing the stresses to their social and economic sources. It then calls for creativity and pragmatism in formulating strategy.

The planning discipline is a sound methodology, and it is grounded in The Nature Conservancy's experience in system conservation. Following it calls at times for a kind of tunnel vision. While tunnel vision is not such a bad thing for moving forward with a project, we should not forget that the project is going to emerge from the tunnel. If we are focused, for example, on the quality of water in the Big Darby and we have decided we can improve that quality fastest and best by promoting alfalfa production, we may stop thinking about the forests of the watershed and focus our energy on meeting the numerous challenges associated with our strategy. But in the long term, forested riverbanks will better secure the river system. In a periodic review of the Big Darby program, we would ask whether the pragmatic alfalfa strategy we are inclined to pursue sells the ideal long-term forest vision we have for the system down the river. If we aren't so aggressively promoting alfalfa that *existing* riparian forest is converted, we will probably find nothing in the alfalfa strategy that is fatally inconsistent with an ideal long-term vision.

System conservation is like a series of journeys that take the planner not to a different place but to different dimensions of the same place. On each trip, some parts of the conservation landscape will have changed because of the work we have done. And some features that haven't changed will appear to have changed in the light of the new perspective we've obtained.

It is that evolving perspective which makes periodic review both useful and necessary. The review required is, in a sense, a renewed inquiry into system definition. We ask again whether in light of the conservation experience gained (or at least with the perspective gained from having completed the first planning cycle) there are elements of the system that should have been included in the definition, that should have become conservation objectives, that were incorrectly ignored, overlooked, or obscured.

One way to facilitate this broad inquiry into the decisions made about system definition is to ask what the natural system ought to look like were there no constraints of cost or need to compromise. With that vision well fixed in mind, we then try to predict what the system will actually look like under the strategy and management prescription contemplated in the conservation plan. That probable future of the system must then be compared with the ideal vision. If there is a substantial difference, the reviewers must examine the reasoning under which the attenuation of the ideal vision was justified. They must satisfy themselves that the compromises and accommodations reflected in the strategies should have been made. They must again, at this point of the process, ask whether they are satisfied with the way the system will operate if the plans they have made are successfully carried out.

The review also offers an opportunity to see whether the strategic solutions envisioned for the system support our broader goals and aspirations for the conservation movement and society. If pragmatism is a useful tool for getting things done, the tool has a second edge. Creative and pragmatic approaches to system conservation can place conservationists in the position of promoting or facilitating activities that compromise the integrity of the system we are trying to conserve. We find ourselves in that position because we are trying to avoid a worse result, the utter destruction of the system, but the cost of following the course of compromise can be high, and we had better be right. It was Oscar Wilde who said, "An idea that is not dangerous is unworthy of being called an idea at all." Our conservation goals require Wildean strategy ideas of us, and because they do, we need to consciously remind ourselves of the long-term results we want to achieve and affirm that the risks we are considering at each stage of the journey are worth it.

Oscar Wilde would have found worthy of the name the ideas about growing alfalfa at the Big Darby, building houses at the Virginia Coast Reserve and Cosumnes, and working with ranchers in the Borderlands. Strategies for preemptive development are dangerous enough that project teams adopting them ought to review them with the greatest objectivity attainable, considering carefully the risks inherent in them for both their focused and their broader conservation goals.

The reviewers must then ask one final question: what will the system look like if not all but only most or even some of the goals of the plan are met? That question is in some ways the one we least like asking, though indeed it was surely asked at least intuitively at the threshold of the project, before the resources necessary to create the plan and take the first steps toward its implementation were committed. It is an unstated question, too, throughout the planning process. It needs to be explicitly put, and answered, at the end of each planning cycle.

When this question is asked, the realist, the businessperson, the handicapper will respond that not everything you plan to get done will get done. That response will prove correct almost every time. This is especially so if the plan is ambitious enough, if it's a major system under real threat and the plan is to make an important series of changes in that situation, to reverse an apparent trend of destruction.

Somewhere within each of us working in conservation, there must be a place for the realist, the handicapper. There are some projects we should abandon and some we should never start. But it is even more important to harbor within ourselves the person who won't take no for an answer, who will go ahead and reach for the barely attainable. Daniel Burnham, a turn-of-the-century Chicago planner who is credited with having had the foresight to envision the Cook County Forest Reserve, is supposed to have said, "Make no little plans."

Many of The Nature Conservancy's most impressive accomplishments can be traced to starts that would have been easy to handicap. The rational bettor would have given big odds against them. Correspondence, for example, in the Conservancy's files regarding the chain of barrier islands at the core of the Virginia Coast Reserve dates back to 1956. The acquisitions that gave the Conservancy a reason to try to save the system were *twenty years* in the making. Indeed, Conservancy lore has it that, apparently stymied in an attempt to acquire an important holding, the organization finally used a wholly owned subsidiary with a name that didn't reveal TNC as the parent corporation to make the critical transaction. If you are flexible and determined, you can change the odds.

Most people would have bet there weren't any sizable remnants of Indiana's native prairie left by 1960. One woman beat those odds by finding such a prairie, and beat even bigger odds by getting it preserved. The odyssey of Irene Herlocher began when a friend with gardening interests showed her a "secret" place about thirty miles south of Lake Michigan where some unusual plants grew. Mrs. Herlocher visited the place with growing fascination over the course of a year or two. At a college reunion, she chanced into a slide presentation on the nearly vanished prairie heritage of the Midwest and watched as many of the plants from the secret place filled the screen. It slowly became clear to her that the secret place was a sizable remnant of Indiana's past. The secret place was a prairie.

There is a network of prairie enthusiasts in the Midwest, and with the passage of a little time and the payment of more than a few long-distance telephone tolls, Mrs. Herlocher got in touch with some of the most active of them. These folks, mostly from Chicago, knew about the site and had, indeed, written Indiana officials about getting it protected. There had been no response from official Indiana. So this great place, this fascinating discovery, remained unprotected, and nobody who could change that situation seemed willing to get involved. Except now there was Irene Herlocher. She made contact with Indiana's Division of Nature Preserves and quickly enlisted its moral support for protection of the site. But the agency didn't have the budget or the means to save the prairie, and the land, right on the edge of Indiana's highly industrialized Calumet region and a block away from a major north-south artery, U.S. Highway 41, was going to be expensive. Mrs. Herlocher went to see county functionaries, who, it turned out, had tried not long before to get federal assistance to buy the land—for a golf course! Mrs. Herlocher wrote letters, recruited supporters, and appealed persistently to politicians, civil servants, educators, and conservation organizations (The Nature Conservancy's first response: "How much money can your committee raise?" Her answer: "What committee?"). She organized a committee. She stayed with the project through dozens of occasions for giving up. And she prevailed in the end when, after ten years of trying, the first and biggest piece of Hoosier Prairie was bought, with more to follow.

It *is* possible to beat the odds. I remember sitting on a late spring day in 1989 in a stark meeting room located in a hangar at the Tulsa airport while the board of trustees of the Conservancy's Oklahoma program tried to decide whether it would take the lead in an attempt to buy one of the best unprotected tracts of tallgrass prairie in the country. It was a $15 million undertaking, and the trustees were hesitant to commit Oklahoma to it. Several broadly experienced and well-connected board members told me they would help save the Oklahoma tallgrass prairie, but they said if they were able to raise $750,000 in Oklahoma they would consider their efforts a major success. A million dollars was probably out of reach; $15 million was out of the question. Their predictions were well considered, they were sympathetic to the cause, and they were the right people to ask. The realist, the businessman, the handicapper would have turned away from the project. I didn't think the Conservancy could pay for the place without a lot more than a million dollars worth of Oklahoma help. (Yes, I have noticed that my response looks more than a little like the one Irene Herlocher got—"How much money can your committee raise?" We usually have to find most of the money required to protect even an area of national significance in that area's

Figure 11-1. Tallgrass prairie near Pawhuska, Oklahoma. © Harvey Payne

own backyard.) There were a couple of people on that Oklahoma board, though, including the chairman, Joe Williams, who were so drawn by the power of the project, so convinced the prairie had to be saved, that they would not turn away. They lobbied for a commitment of national support. They pleaded, cajoled, inspired, praised, and planned, nationally and locally. About $10 million was raised over the next four years in Oklahoma for the prairie. With that kind of unimaginable local support, the Conservancy managed to raise the rest of the money from national sources, and a prairie of more than 30,000 acres, big enough to host its own herd of bison, joined the ranks of protected areas in the United States.

*   *   *

Set clear goals and establish clear measurements to track your progress in pursuing an ecosystem approach to conservation of a natural system. Make good, realistic assessments of what will happen if all the goals in your conservation plan are not met. But don't let the results of that analysis drown out the message from the other side of the brain. Get the conservation done because it is important enough, powerful enough, inspiring enough that it *has* to get done. And with respect to the odds, remember that if there was no reason to think that (with apologies to Lincoln) the better angels of nature were going to intervene on your side, you wouldn't even have started.

PART III

Ecosystem Conservation in Context

CHAPTER 12

Props in the Set

Most of the questions this book was written to answer have to do with making the choices that arise in an effort to conserve a site, a place on the landscape. As a consequence, the scope of many of the issues discussed has been narrow, and the resolution more or less within the control of the folks working on the conservation projects. With respect to other issues that have been mentioned, the conservation team's posture, if not dispositive, could be expected to make a substantial difference.

But site-based conservation obviously exists within a larger context. Within that larger world, there are many forces which a site-oriented conservation team is not likely, acting alone, to be able to exert a strong influence. No discussion of system conservation would be complete without some reference to those factors, and literacy in the issues surrounding those factors is required of system conservationists even if it isn't immediately clear what might be done about them, if, indeed, there is anything that a system conservation can or should do. Global climate change and population pressure are two special cases that have been discussed, but there are numerous others— war is, too frequently, an example—that can affect or destroy a conservation site but that are not likely to be much influenced by system conservation concerns and efforts. International markets for produce and natural resource–related materials are not quite so inert. One can more easily imagine successfully attempting to influence global commodity prices from a conservation base through national legislation or massive public relations campaigns (the coconut oil market was affected, for example, by emerging consciousness of cholesterol as a health concern), but the gold and tin markets, and even the banana and coffee markets, are huge and are governed by a complex set of influences not often superseded by work conservationists are able to do. National legal policy is somewhat less daunting but still a challenge. Like all these issues, it represents a kind of larger set in which the conservation drama is played. Change a prop in the set and the show takes on a different meaning, the script produces a different effect.

Legal policy is the product of many influences beyond those originating at the typical conservation site. There are, however, likely to be occasions for system conservationists to address legislators, voters, or the news media on national legal issues, and we should rise to some such occasions. No book of general applicability can by itself adequately prepare system conservationists to fulfill that obligation in a specific instance. But it is essential that we understand some broad themes relating to environmental law and society. That understanding will serve as a foundation for thinking about specific issues.

Environmental regulation, like all law, is designed to produce or protect something society values while distributing the burden of doing so in a way society perceives as equitable. Doing the things that will produce or protect environmental values and that will prepare us to prosper in the long term often carries some cost in the short term. We have been conducting our business for a long time, in much of the world, as if there are alternatives to paying those costs. A course that has seemed to us an alternative has been to go over the next hill when we wear out the land we're living on now. That, however, doesn't really permit us to avoid the cost; it just delays the date of collection. We've now been over most of the hills. The long run is here, to squarely answer the well-known Keynes comment quoted earlier in the book, and we aren't dead. We have instead left a series of unpaid debts behind us—we have exhausted the soil here, fouled the water there, destroyed the ecological balance over there—and we have arrived at the time of payment: of adjusting our activities because we can no longer plan on settling over a new hill but must instead plan to preserve the integrity and productivity of the places where we are. The first order of business in a lot of these planning venues is to deal with old debt, the debt of old ways and old events; there is a payment to be made, in dollars, in cleanup, in changing the way business is done. Where there is a payment, there is a question: who pays?

While some of the costs of protecting some environmental values have been specifically allocated in existing laws, society's most frequent answer to the question of who pays is to continue to deny that it is time for payment. That is itself, of course, an answer to the question of who pays: future generations. Some of us have comforted ourselves with the notion that the cost is slight, the loss of a few species trivial. Perhaps it is useless, as the African proverb suggests, to try to awaken people who are pretending to be asleep. But it isn't just a few that we're losing, and it isn't just species. The reason, indeed, that it has become necessary to explore ecosystem approaches to conservation is that we need to find ways to stanch the loss of large parts of the functioning natural world that supports species, including our own. It is time for payment, for acquiring land to preserve natural systems, for replacing exploitive businesses with compatible ones, and for replacing government incentives for development that ignore environmental consequences with incentives to conserve. Certainly the question of who pays is a moral one, but if we are going to secure the needed payment, we have to consider it a political one as well.

One way to sort out the political and moral questions is to ask what's fair. What we think is fair is governed, in part, by what we think is important. In some ways, that's a much more critical question than what the law is: if society does not believe a thing is important, it will come to treat a law that protects that thing with contempt. If society comes to believe something is important, its legal system will eventually reflect that belief by protecting that thing. Indeed, traditional communities with particularly clear dependencies on identifiable natural resources often develop social understandings that prevent overexploitation without the formality of law. A fisheries-dependent community might, for example, customarily restrain its harvest at seasonally important fish aggregation points. Such understandings are enforced with powerful social sanctions. New social factors—such as the development of a cash economy—can throw conservation understandings out of balance, but their existence is nonetheless evidence of the capacity of our species to reach such social agreements. In any event, a lot of environmental problems now and even more, in all probability, in the future will have to be resolved without reference to legal mandate. Under an ecosystem approach to conservation, that is nearly always the case on at least some issues. Because there is no legal framework defining or regulating ecosystems or ecological processes, many of the strategies discussed in chapter 8 were designed to facilitate, without legal mandate, a transformation of human economic activity that affects natural systems. Such change doesn't happen quickly. The Nature Conservancy's work at Pyramid Lake–Stillwater Marsh in Nevada illustrates both the kind of environmental challenge that requires transforming conservation and a private and public response that has begun to address such a challenge in the absence of responsive regulation.

Stillwater Marsh was the terminus of the drainage basin of the Carson River and a critical stopover for shorebirds on the Pacific flyway. A hundred years or so ago, it appeared to some that the desert environment west of the marsh would be a fine place to grow alfalfa if the water that fed the marsh could be diverted to that more remunerative purpose. The flow of the Carson alone—the water that had sustained the marsh—was deemed not robust enough to satisfy the needs of the farms-to-be (much less the town of Fallon, which grew around them). Pyramid Lake, fed by the Truckee River and previously unconnected to the marsh hydrologically, was plumbed into the mix with a canal. And so people started to grow alfalfa in the desert—

pretty tough on the marsh, which got smaller and smaller and saltier and saltier. Because less and lower-quality water reached Pyramid Lake, it wasn't so great for lake fauna either, particularly a fish, now listed as endangered, called the cui-ui. The Paiute Nation, which has treaty rights to fish in Pyramid Lake, has in recent years been moved to respond according to a European American tradition that is despised in rhetoric and honored in practice—it sued.

Undoubtedly, it would have been far better environmentally, and more sensible economically, to have left the water systems of the Truckee and Carson basins alone in the first place. We didn't. Much of community life in Fallon is built on decisions made long ago to dry the marsh and compromise the lake to support an agricultural effort that nature didn't mean to be there. Gains from resuscitation of the Stillwater and preservation of Pyramid will be shared by everyone who cares about the natural world—a critical marsh used by migratory birds could be restored, an endangered species secured. The economic and social loss of turning the system back toward the form in which it was naturally engineered, however, must fall in unequal part on the residents of Fallon. Aside from the question of whether leaving it there would be a winning political strategy—it usually isn't—the fairness of a decision to distribute the burden in that way ought to be questioned.

It is true that the local populace gained the most from the decision to exploit and exhaust the natural system in the first place, but most of those gains were received by folks long gone. What's left is a big environmental debt that was generated over a long period of time. Somebody has to pay. The payment schedule we advocate needs to account for the current residents' political ability to delay payment further, but not only that. Ecological advocacy supports a general and long-term interest. If that interest is to prevail against a short-term interest that is acute and specific, the manner in which the ecological interest is advocated must be perceived as fair, and our compassion with respect to the question of how payment is to be made must be evident.

Working at Stillwater with the Environmental Defense Fund, the Conservancy advocated and carried out a series of purchases of water rights owned by some alfalfa irrigators and sold the water it acquired to the U.S. Fish and Wildlife Service. The Service will leave that water in the Carson and Truckee Rivers, to the eventual benefit of the lake (gaining the parties a reprieve in the Paiute lawsuit) and the marsh. That costs many of us something as federal taxpayers. But the community also pays. The individual community members selling the water rights are directly compensated, but the community as a whole will have to deal with changes in its life and economy. The fascinating if overused dissection of the Chinese character for crisis applies here. There is danger in the community's position, and there is

Figure 12-1. Irrigated pasture near the Carson River, Fallon, Nevada. © Daniel D'Agostini

opportunity. The disputes over water in the Truckee and Carson basins are far from resolved. Fallon has, however, received a signal that is more gentle than that which would have eventually been received from another source. The message is that it is time to initiate a process of transition to a fundamentally more secure, sustainable community, a community not dependent on more water than nature usually delivers in its watershed. The Conservancy has begun to work with county leaders to explore the implications. There probably isn't a fairer overall resolution to be crafted.

Fundamentally, fairness must mean that the full spectrum of reasonable concerns regarding land use and the direction of society and development will be considered when decisions are made. Some examples of how that might be done have been recounted in this book. I believe that there is still a need to do better in finding ways to bridge the gap between the concerns of those who want to make the most money in the shortest period of time and the concerns of those who want most to pass on a rich, diverse, and healthy environment to their children. If we do nothing about that bridge, the second group will often be left on a barren shore.

The quickest way to bridge the gap—to make the payments needed—is to use money to do it.

Government, when it is working, exists as a vehicle for the expression of collective will. If society has the will, its government can collect ample resources for the necessary payments toward setting the nation on a sounder ecological foundation. We could make a good start by redirecting some of the payments we already make. Government has not generally directed payments toward inducing payees to employ sound environmental practices. But we could, for example, pay farmers to use best practices to save topsoil instead of guaranteeing a minimum price for corn and soybeans. We could expand programs that provide compensation for taking marginal and vulnerable land out of agricultural production. There are social and cultural reasons that add to the environmental justification for supporting farmers with these kinds of subsidies—family farmers have been prominent threads in the American business and social fabric, and the tradition of family farming is unraveling under a variety of pressures. Environmental subsidy programs need not even be permanent—they could be planned to facilitate a ten- or fifteen-year transition toward an agricultural economy that operated in freer markets and under higher environmental standards as a matter of habit.

With the federal budget already oversubscribed, however, and state and local budgets under similar pressure, society may not have the will to create new programs even if doing so would help the nation achieve objectives its people consider important. If we fail to create new programs, are we therefore bound to tolerate environmentally irresponsible behavior? Many who think our society is overregulated have argued that society cannot legally constrain their behavior with respect to the environment without paying them. They cite as authority the Fifth Amendment of the United States Constitution, which provides that private property shall not be taken for public purposes without just compensation.

To respond in short, subsidizing environmentally responsible behavior may be good policy, but the absence of such subsidy payments does not render society unable to proscribe environmentally destructive behavior. As long as there have been laws, diminution in private property values has accompanied their enforcement. Laws make it possible for people to live together in communities in tolerable harmony. Laws also constrain individual liberty. The Constitution prescribes, in general terms, a set of limits on the compromises to liberty that individuals must accept in return for the benefits and privileges of living in community with others. With respect to our liberty of property, and specifically with respect to property not physically appropriated, the Fifth Amendment requires that a property owner be compensated when a government action eliminates essentially all, or at least a very large part of, the value of that owner's property.

When the U.S. Supreme Court considered the constitutional status of zoning upon a Fifth Amendment challenge, it proceeded by comparing individual property rights with the right of a community to shape its character with regulation that constrains individual rights. Land use regulation laws are the principal means by which communities can shape their character. Laws and associated regulation that have had the effect of diminishing the value of property by as much as 90 percent without providing compensation have consistently been ruled constitutional by the Supreme Court of the United States (see, for example, *Penn Central Transportation Co. v. New York City,* 438 U.S. 104 (1978). Those who doubt that the founding fathers of the nation would have found such a result acceptable must explain how the courts could have rejected all claims for indirect taking of property by regulation during the first hundred years of Fifth Amendment jurisprudence. On the whole we have, as a society, broadly accepted the incidental burdens that all laws place on property rights. We have accepted those burdens as the price of admission to the community's dome of warmth, comfort, specialization, and protection. Indeed, the burden is so familiar that we don't often think of it as such when it comes to most laws. We may not always observe speed limits, but we accept them well enough that we have stopped arguing about whether they cause an unconstitutional diminution of the value of our cars.

Environmental regulations such as those specifically designed to protect wetlands and endangered species are among the newer burdens local and national communities have decided they must require members to bear, and society is still sorting through its feelings about accepting those burdens. Acknowledging that the fairly well settled burden of zoning (highly controversial when the first ordinances appeared) is distinguishable at some level because it is a regulatory scheme imposed at the option of local government, the Fifth Amendment analysis of the property rights burden inherent in environmental regulation unfolds in a manner quite like that which prevailed long ago when zoning was being questioned on similar grounds. Like zoning, environmental laws impose incidental burdens on property to accomplish legitimate societal purposes, and unless they eliminate substantially all of the value of affected properties, they surely do not run afoul of the Fifth Amendment. Critics of environmental laws might respond that they have no love for zoning either, and that the plain meaning of the words of the Fifth Amendment makes it clear that any court presuming to rule zoning constitutional is wrong. But it is often not plain what a few hallowed words mean when they must be tested in one of the infinite variety of disputes in which they might be relevant. The meaning of the Constitution for legal purposes cannot practicably be left to each individual's judgment and discretion. There must be a final arbiter; that arbiter is the United States Supreme Court, and the Court has not read the plain words to mean that any government action–induced diminution in private property values requires compensation. It is hard to imagine, moreover, what alternative to zoning a community could constitutionally employ as a tool for maintaining what most of its members see as a reasonable quality of life had the Supreme Court reached a different decision on zoning.

Besides, if the real issue for those segments of society that complain about the burden of environmental laws were the taking of private property for public purposes without just compensation, the debate could not stop with environmental laws and would have to address a variety of laws that don't seem to be at issue: safety regulation, child labor laws, and the speed limits cited earlier as having become second nature to us. The debate isn't really about the Constitution or about what level of government makes the laws. The issue, at base, is not the constitutional status but the appropriateness of environmental laws, or, to put it another way, whether or not the time has come for property owners to pay some environmental debts.

One reason there is disagreement on that subject is that events have rapidly changed the way society must regard property ownership, and a lot of people haven't managed to keep up. A certain amount of destruction or disruption of nature has always been accepted and regarded as no more than an incident of reasonably responsible tenure in the property. We expect to be able to walk on our land, though our footsteps destroy insects. We expect to be able to replace plants and soil on a part of our land with the foundation for a house. We don't expect to be criticized for planting a suitable crop or husbanding a suitable number of livestock. Within our legal structure, we expect to be able to do these things because we own the land; our ownership has been thought, by implication at least, to extend to the natural features and organisms we destroy or disrupt when we exercise the prerogatives of ownership.

But the range of things we once could reasonably have expected to do on our property has been restricted by events. The imminent danger of extinction of so many plants and animals is one of the change-inducing events. Our concepts of appropriate land stewardship and property rights were developed in the absence of an extinction crisis. There has been a societal response to that crisis: the legislation of new standards for land stewardship in a variety of environmental laws, including the Endangered Species Act. These laws imply an answer in the negative to a question of property law that did not exist as our old concepts of property rights evolved and that could hardly have occurred to the framers of the Constitution: does a landowner have a property right in an action that destroys the last living member of a species?

Events have changed the nature of property law even more broadly: the existence of so many people with so many demands and so much technological capability to fulfill them has profoundly changed the context for the application of such venerable principles as *Sic utere tuo ut alienum non laedas*—"One cannot use one's property in such a way that it damages the property of others." Indeed, much environmental legislation can be thought of as a societal acknowledgment of the new implications of that familiar principle of law.

For example, as a society we at first applauded, and then treated as part of the general order of things, the destruction of the first 50 percent of the wetlands in the continental United States. If asked, we would have said swampland owners were making wasteland productive through drainage. Events have changed our understanding of swamp drainage. We have begun to understand as a society that those who are destroying the wetlands that yet remain in this country are hurting the rest of us. A thoughtful application of the common-law maxim *Sic utere tuo* to the issue of wetlands destruction now must hold that swamp owners don't have the right to use their property in that way—that draining swamps means something different now from what it once meant to society. Generally, further loss of wetlands costs society money in flood damage to other property. It makes other private property less valuable for watching or hunting the birds that nested and found refuge in the wetlands that have been destroyed. It makes our boats less valuable by destroying the spawning habitat of fish we liked to catch. It makes it more expensive for us to have clean water. Property owners who drain or fill their wetlands take public and private property for private purposes without providing anybody any compensation. Draining of wetlands can no longer be a right that accompanies the ownership of property. Events have changed the nature of wetland ownership.

When our legislators were finally moved to pass laws designed to slow down and sometimes stop property owners who wanted to drain and fill wetlands (codifying and supplementing the permit requirements of the Rivers and Harbors Act [1899] in section 404 of the Clean Water Act [1972]), society therefore wasn't taking anything away from anybody. Because there are more people in the country than ever before, because we have better transportation than ever before, and because we have the capability to alter the landscape to make more places suitable for development than ever before, there are more times now—a lot more—when doing whatever we want to do with our property hurts somebody else. Now would be a good time to start getting used to the limitations that new reality implies. The environmental laws that now exist are evidence of a dawning societal recognition of that reality. The controversy such laws have engendered should not be too surprising—recognition will not dawn on everyone at the same time, and the response the new reality requires is not trivial in scope. The Endangered Species Act, viewed in one light, requires that a property owner govern activity on his, her, or its property such that the property remains suitable habitat for certain species. That is a new kind of responsibility, and it is taking some property owners time to adjust to the idea. We can be sympathetic, but the legal foundation for an argument that they should not have to make the adjustment is, at its base, startlingly unpersuasive. The aggrieved property owners must argue that there exists some legal protection for an owner's prerogative to use land in a manner that causes or contributes to the extinction of a species.

Admittedly, there is little in the history of property law that provides guidance. There haven't ever before been enough endangered species to support the development of a body of common law on that subject. There is a long history, however, of legal limitations on property rights with respect to wildlife generally. The force and effect of that law is that wildlife is owned by the state. In English law, from which most of our law is derived, wildlife was owned by the Crown. In the United States, courts have ruled that the people stand in the stead of the sovereign and can, through their government, make laws that regulate or prohibit the taking of wildlife.

But by regulating habitat destruction, the Endangered Species Act moved well beyond traditional wildlife law. Destroying habitat is sometimes an incidental effect of activity that enhances short-term economic value. It has been argued that a law that mandates forbearance in such a case is a taking that requires compensation under the Fifth Amendment. That can be so, however, only if the property rights of landowners in the United States for constitutional purposes include the right to cause extinction. If there is no such right, then a law prohibiting acts that cause or contribute to extinction can give rise to no constitutional claim.

I don't think that property rights, constitutional or otherwise, can reasonably be thought to extend to the destruction of a species. Endangered Species Act or no, in the end it is a question of values or morals. The continued existence on earth of a species is an asset that cannot be held to belong to one person. If humanity can even be said to have an ownership relationship with an entire species, ownership must belong to all humankind, past, present, and future. Eliminate from the planet the unique set of common characteristics that make a group of organisms a species—the Carolina parakeet, the ivory-billed woodpecker—and you kill history, deprive your neighbors, and rob your children. It cannot be within the prerogatives of an owner of land to use it in that fashion.

Just as the loss of wetlands has emerged as a clearer issue with every day's thousand acres drained, actions that once were not weighted with extinction implications now are so burdened. But that sad fact doesn't change a property owner's position with respect to engaging in species-destroying actions or support a claimed right of compensation in return for forbearance. Our community has grown in numbers, in power, and in destructive legacy. Development that was once progress is now robbery. There was never a protectable property interest in robbery.

If we can be firm about where there is and is not entitlement, we can be generous in offering assistance to those who must bear the cost of change in the nature of property rights. Pollster Celinda Lake has concluded from her opinion research that environmental goals would be better served if, when we insist on changes in behavior for environmental reasons, we adjoin to that insistence a commitment to assist in the process of transition. She says there is widespread support in the United States for environmental laws. Most people want to encourage environmental protection, and they will eventually demand it. But there is also, Lake says, genuine concern in our society for those who are hurt in the transition toward a sound relationship between our society and our environment.

In that spirit, we ought to consider the hardship that might be imposed on an individual or family or business when a surprise endangered species habitat issue arises. I think society would be acting fairly if it decided to indemnify from capital loss those property owners who bought land before a species was listed, held it for five years, and found on sale or transfer that endangered species–related restrictions, as distinct from market conditions or other measurable explanations, caused a capital loss. But the reason for making any such assistance available is important. It is social assistance with the kind of transition Lake was describing, not compensation for a property right taken.

The topic of public environmental policy merits and has received treatments that would fill shelves, and it will not be discussed much more here. This brief bow to it has focused heavily on the social foundation for a couple of new environmental laws. The treatment is therefore unbalanced, and it must be righted: the weight of public policy is not expressed in its new and relatively few environmental laws. The balance of public policy is heavily against a healthy environment, against conserving reasonable integrity in our nation's ecological systems. From the way we keep our national accounts— we don't acknowledge the use of natural capital as source material or waste receptacle—to our subsidization of automobile travel, to the weight and effect of our farm programs, to a long string of tax law–inspired building and development fiascoes, the long odds against success in system conservation are made longer yet by the loaded dice of public policy. Environmental laws have merely mitigated that overall policy effect somewhat with respect to some of the most ecologically important land in the country. There is every reason to be resolute in defending the laws we have. If we waver, we won't avoid payment; we'll just delay it a little longer or add to the debt we pass on to our children.

We can make plans for payment that acknowledge the difficulty, that ease the transition. Some of the best examples of fair resolution of tough environmental conflicts have emerged from Habitat Conservation Plans (HCPs) under the Endangered Species Act. HCP provisions make it possible for representatives of the government to meet with owners of land that is habitat for endangered species, to consider the full range of options for conservation, and to make a comprehensive plan designed to save the species and accommodate development interests by protecting certain agreed-on habitat, while releasing other habitat from regulation under the Endangered Species Act. Securing the habitat to be protected usually costs money, and part of the HCP negotiation process has involved arriving at an equitable and practicable formula for sharing that burden among developers and local, state, and federal governments. The agreements that have been struck have found a balance that typically includes a substantial role, both for those who want to use endangered species habitat for economic gain, and for the taxpayers, for whom the benefit is avoiding more irretrievable losses of a natural heritage. That seems pretty fair. The Conservancy has been deeply involved in the negotiation of several of these plans and has worked effectively with developers in those instances. Some of those developers might even have talked to the Conservancy absent an Endangered Species Act. But as a foundation for discussion, the act is critical. Without the fire of the act under the crucible in which those negotiations took place, no new compounds would have emerged. If we waver—if, for example, society finds the burden of environmental responsibility too heavy and purports to determine that developers have some kind of property right in driving species to extinction—the negotiating stances will change so much that we may never see another Habitat Conservation Plan.

There is a Native American tradition about planning for the seventh generation. Who knows what the compounding environmental debt will cost over time? But assume that a modest interest rate of 8 percent per annum reflects the costs to society, starting today, when we destroy another thousand dollars worth of habitat or aquifer. If the debt incurred today comes due in seven generations, if then finally, the time for payment cannot be delayed any longer, our grandchildren's grandchildren will have to come up with $47 million.

Considered as an economist would consider it, the figure isn't so startling. A thousand dollars today equals $47 million in 140 years at 8 percent. Cost today plus interest equals cost tomorrow. Period. Any inclination we have to worry about the environmental debt we bequeath to future generations based on its compounded cost, is driven more by emotion than reason; unless there is something rational as well as spiritual about our concern for the environment that can not be fully measured in dollars today and dollars tomorrow, or unless perhaps, the various environmental debts we are incurring also compound each other.

Maybe there is something rational in our emotional inclination to begin to worry about what will happen when our environmental debts come due.

CHAPTER 13

Beyond the System

Several years ago, Jose Lutzenberger, formerly environment minister of Brazil, addressed a group of conservationists, international aid officials, and United States legislators at a gathering organized to celebrate the launching of an initiative called Parks in Peril. In much of the world, the designation of a park is an important but not definitive step and is far from a guarantee of reasonable protection. The Nature Conservancy designed the Parks in Peril initiative to make at least eighty important Latin American officially protected areas into something more than "paper parks" by providing resources for basic management.

Following congratulatory remarks offered by an Inter-American Development Bank official and a spokesman for the Agency for International Development, Lutzenberger approached the microphone. He wore a black suit. The front collar of the white shirt he wore under a black sweater lent a clerical impression to an appearance otherwise dominated by the slightly preoccupied manner and unmanageable gray hair of the stereotypical scientist. Lutzenberger moved quickly through an obligatory comment or two about how Parks in Peril was an important program and a necessary step in the right direction. His voice then rose; he wasn't preoccupied now: "But isn't the very notion of parks an obscenity!" The words just tumbled out. Brazil's environment minister took offense at the notion that the world had been reduced to defending a few "green measles" on the map as our hope for an environmental future. He urged that we rethink and reform the ways we are living, and he singled out the United States for special mention—Americans' addiction to automobiles struck him as especially egregious. Unless people begin to live more conscious of ecological limits, he said, the green measles will be our reminder of a diseased world and our anemic attempts at a cure.

Whether land conservation efforts are addressed to particular species, to small natural areas, or to larger landscapes and waterscapes, they are part of an overall strategy for conservation of the world's biodiversity that is, at this point, failing. Europe and the Near East have been human-dominated landscapes for a thousand years or more. Much, perhaps most, of the battle is already lost there. Much of the natural integrity of North America has been lost in the several hundred years during which Europeans have spread themselves across the continent, and the really large wild areas of the rest of the world are fast being tamed. What's left ought to be saved. But if existing site-based conservation efforts aren't complemented by efforts to influence the market-driven land use practices that affect lands beyond those we formally and specifically protect, we will lose most of what is left.

To some, perhaps especially to those who have seen how impressively the conservation movement has grown and how difficult it has been to accomplish what has been accomplished, that news may come as a surprise; our relative success tends to delude us into thinking we must be winning on an absolute scale. Then too, many of us have seen the various "official" biodiversity strategies, which suggest implicitly that with good planning and management and the provision of an ambitious but reachable level of resources, the biodiversity of this nation or that can be satisfactorily preserved in a relatively small amount of parkland.

To others, it isn't news at all. Not a few of those have given up the fight as a lost cause. One of the saddest descriptions of the conservation community's work I have seen was published in *Nature Conservancy,* TNC's magazine. A Conservancy officer described the biodiversity conservation challenge that remains before us as effecting an orderly retreat: a remarkable statement to make without elaboration, a sentence of doom. We can hope for more than that; indeed, we must plan for more than that. We owe it to future generations to accomplish more than that.

The site-centered planning methodology presented in this book will be very useful for the sites to which it is applied, but to reiterate a point already made, it won't by itself enable us to reverse the orderly retreat. If system conservationists pay attention where we work to developing solutions that will help others in other places develop solutions, we have a chance to slow the retreat, to establish a defensible line, and perhaps to reclaim some lost ground. Where must that line be drawn? How much of the landscape must operate with how much integrity to preserve a satisfactory whole? How much can we lose before no reasonable recovery is possible? These are hard questions to which there are no definitive answers. There have, in fact, been relatively few attempts to find answers. But the questions are useful ones to think about, and a good place to begin thinking is with the current situation on the land as a whole.

It may be useful to expand a frequently used metaphor and imagine nature as a physical and biological net that supports and is partly constructed of a wonderful multitude of webs of living organisms. The strands of the webs in which humans have become enmeshed are increasingly numerous. At this point in our development, humanity might even justifiably adopt a kind of pre-Copernican view and place itself in the center of the net. We didn't start there, and indeed, we established that position not by achieving deep understanding of the net but by clumsily and at great cost to the structural integrity of the net manipulating its webs until we ended up at the center. Our strongholds have been established at the cost of weakening the net, and the net is weakest of all right beneath our strongholds. On the lands we use intensively—the urban and suburban areas, most crop and evergreen tree farms, and lands devoted to industrial and commercial development— we have eradicated most native species. These lands not only do not contribute very much to the natural function of the globe, they tend to export biodiversity-stressing products and by-products. Part of any strategy to establish a defensible line, to halt the retreat, must be the maintenance of the legislative and regulatory framework that mandates minimization of harmful outputs from our strongholds. We must also seek new tools, including market- and incentive-based strategies, to further reduce pollution and other so-called externalities.

Also supported by the net but increasingly marginalized are conservation lands. The now-familiar idea that some parts of the net must be set aside to ensure that there are places in which natural processes are preserved is actually a profoundly new one. Until very recently, setting lands aside to ensure that they would be natural would have been superfluous; most of the land on the globe was in a natural condition. The ascent, such as it is, of humankind changed that. There were terrestrial species on earth for several hundred million years before *Homo sapiens* became one of them a couple hundred thousand years ago. The ancient globe must have scarcely awakened to the presence of a new and modestly successful bipedal creature as, 10,000 years ago, it experienced the most recent geological event of continental scale—the return to the far north of the last big glaciers. The retreat of the glaciers opened the field to new landscape-altering events of continental scale. These weren't geological in origin. They were the handiwork of humankind, and the effects of that work have increasingly characterized the globe. Now, when we are seeking strong nodes in the net, places in which the natural structure of the planet is predominant, we have to look for them; they are very special places, and their protection is far from superfluous.

Some such places have been set aside as parks and preserves. Many of those places are also compromised; in and around some national and state parks, for example, because of accommodations for visitors, the strong nodes of the net are a little less strong. Wildlife refuges accommodate hunting and other development, state and national forests accommodate timber harvesting, and other sparsely developed public and private lands accommodate a variety of other development, but contribute even so, in varying degrees to biodiversity and the reasonable functioning of the natural world—the maintenance of the net.

According to a 1990 United Nations report, conservation lands now occupy about 4 percent of the terrestrial area of the earth. That figure is probably no higher than 10 percent in any nation; in many industrialized countries the figure is far lower than that. Russians, for example, take pride in the exceptionally strict system of national scientific reserves that occupies 1.5 percent of the land in their vast country. Russia also numbers provincially managed conservation lands in its park and reserve portfolio; the management of these lands varies in about the same way that management of our national forest, Bureau of Land Management, or state game and fish areas does. It is sometimes said that 25 percent of Costa Rica is in a protected status, but much of the land so classified is managed for timber; far less than 25 percent of the country is well enough protected to reinforce the net that supports life in Central America and, more broadly, the Western Hemisphere.

The point is that if we cannot hope to look beyond a system of parks, reserves, refuges, and protected areas for preservation of biodiversity, ecological processes, and balanced natural resource conservation, we ought to join the Conservancy official previously quoted and concern ourselves with planning an orderly retreat. Protected areas—the natural strongholds of the net—could perhaps amount to 15 percent of the landscape if we make much-needed heroic efforts to make that happen. The fate of the net will, however, be determined by what happens to the remaining 85 percent of the landscape.

To look at a specific case more closely, of land in the United States, something between a quarter and a third has already been devoted to land uses that are intensive and contribute little to biodiversity conservation. It may or may not be society's will, but it is within our means to set aside 15 percent of the country for nature (between 3 percent and 10 percent of the country is currently so managed, depending on the definition used). It will take a great deal of effort to double the country's portfolio of parks and preserves. We should do it—but doing so will not discharge our obligation to the net. To secure the nation's native biodiversity and ecological foundation and to contribute our due to that of the hemisphere and the globe, we must manage our activities on the lands that aren't preserved and aren't yet intensively developed to ensure that those lands—the 55 percent of our landscape that is at present neither intensively used nor carefully preserved—contribute to the security of the ecological whole. Those relatively undeveloped, seminatural lands and waters stabilize the whole system of webs. They do that by stabilizing individual webs, the plant and animal associations that exist within a landscape context, securing the processes of pollination and seed dispersal as well as microtemperature, wind speed, and local humidity in air and soil. They stabilize the net as well by securing larger contextual processes: the generation and regulation of climate, groundwater recharge, surface water management, and the chemistry of the atmosphere. They affect, moreover, innumerable other named and unnamed natural processes. The consequences of the diminution and isolation in fragments of the natural landscape have been considered in island biogeography work, but we have spent less time trying to understand the degree to which a seminatural surrounding landscape minimizes loss of biodiversity from core natural areas. We must answer, too, an even more cogent question: what degree of compromise will that landscape accommodate and still play that stabilizing role?

It is exceedingly difficult to establish in advance the point at which the progressive displacement of the natural landscape by human-created alternatives will so fray the ecological cords that bind together the remaining natural elements that the cords will snap and the natural community disintegrate. It may be that migrating birds are the canaries of the world-mine; many populations are seriously diminished, and while the decline in some species is traceable to loss of specific breeding or wintering habitat, there is no such clear explanation for the decline of others. It may indeed be that increasing percentages of usable habitat have been lost everywhere birds need it, and the loss of seemingly nondescript lands along migratory corridors may be as devastating as the loss of avian gathering places. We would be prudent to pay attention to the decline in population of some bird species as we consider the costs of additional habitat destruction. The ecological cords, certainly weakened, may in fact already be dangerously frayed.

Recent research provides evidence that whether or not birds need a lot of relatively natural habitat, the forested landscape, at least, needs birds. In one study, white oak trees that were wrapped in net so as to admit insects but deny access to birds suffered about twice the insect depredation of control trees, and they produced fewer leaves in following years. The research result isn't surprising, but its quantification of the interrelationships of species and natural community is important—and the results suggest, more broadly, that the loss of relative natural integrity in the landscape is likely to have broad, destabilizing, unpredictable, and expensive consequences. Our best hope for preserving a functioning net and the net-dependent webs that are critical to our survival lies in exploring the extent to which the most essential contributions of the seminatural landscape can be maintained while the landscape is used—as critical parts of the net will always be used—to generate economic gain.

It has, therefore, become critically important that we develop economic uses that occupy but do not destroy the landscape and that we launch businesses under which such economic uses can be organized. Those businesses must provide enough of what society wants from, for example, the 55 percent of land in the United States that fits into neither conserved nor intensively developed categories. The goods society wants must be produced and delivered without fatally compromising the integrity of the strands and webs of the net that are connected to the nation's seminatural landscape. Success in managing production within these constraints will require information, commitment, and luck. Success is crucial because failure—fatal compromise—not only will destroy the middle landscape in question; it will ultimately destroy the net.

Nobody wants to destroy the net. Our problem as a society is simply that we think we want other things more, and more urgently, than we want to take care of the environment in our quest to get them. There is notable demand for ideas that offer a way around that conflict, some hint about how desired economic ends can be achieved while preserving the ecological foundations of the environment. We are gaining experience right now with businesses that meet that goal. Very probably, such businesses will feature new ways to pursue familiar themes. There will be residential development, but it will be of a lower density and will employ more sensitive siting than is now usually seen. There will be grazing, but animals will be moved more frequently and ecological health will be monitored more systematically than is now the usual practice. There will be ecologically compatible forestry, alternative crops, and alternative ways to grow and market familiar crops.

On the whole, compatible businesses will demand less of the natural resources of the environment in which they exist. Their chances of sustaining themselves will be enhanced if they can command a premium price for their products. Market research consistently shows that consumers are willing to pay environmental premiums for at least some products, though the characteristics, preferences, and limits of that willingness are yet to be tested through experience.

Environmental premiums are, in any event, only one facet of a multifaceted phenomenon that is at once nascent and widespread. People have begun to realize that they can best achieve a high quality of life if they address economic needs and conservation priorities simultaneously. The relatively few existing efforts that are even tenuously integrated receive a surprising amount of attention. Somehow, for example, a copy of a Conservancy strategic plan that addresses both conservation and development needs of the Clinch River watershed of Virginia and Tennessee got into the hands of a team of northern Minnesota officials with an analogous planning problem. No one from the Conservancy supplied the plan, but the integrated nature of the basic economic and ecological analysis it contained apparently interested some chain of people enough that the plan made its way from southwestern Virginia to the upper Midwest.

Likewise, a Conservancy board member, called to consult with a Canadian provincial government as it tried to make sense of a new constitutional mandate of sustainable development, was surprised to have provincial officials cite TNC's Virginia Eastern Shore work as among the best existing attempts to integrate conservation and sustainable development themes. TNC had at the time (the project has since received several conservation awards) published only fragmentary accounts of that work, and those had appeared in magazines and reports circulated only to TNC members. Word of the project had somehow reached the Canadians, and they were interested in the balance that had been struck at the Eastern Shore between use and preservation. They were interested, as well, in the holistic view the Northampton Economic Forum had taken of resource and development issues.

The reason that the Clinch plan reached the Midwest and the Eastern Shore work interested Canadians—and the reason people all over the United States are interested in Willapa Bay, Chattanooga, and the Malpai Borderlands Group—is that having decided we should do it, we are as a society still looking for the tools and approaches that will *enable* us to simultaneously satisfy the demand for development and the need for conservation. Because people are looking, there is an opportunity for a good site-based conservation project to achieve impact beyond its borders. We can increase the likelihood of achieving such leverage through dissemination of conservation learning if we consciously prepare for and work at it. The commitment and the means to capture and distribute lessons learned must be an integral part of the work plan of a system conservation project from the start.

The learning to be shared from a site-based conservation project will not often, however, take the form of a recipe that will enable those who follow it to stamp out copies across the landscape like so many green cookies. The lessons disseminated will be conservation stories that will, when they are told in a new place, spark a conservation reaction that is unique and that emerges from the particular circumstances of that place.

Some New Mexico ranchers from north of the Gray Ranch, for example, heard that the Animas Foundation, which now owns and manages the Gray Ranch, was making part of the ranch available to host the herds of neighbors who needed to rest range on their ranches. At first, the northern ranchers thought maybe they could use the Gray, too. They came down to see the project and to talk with representatives of the Animas Foundation. Some were interested even after they understood that commitments not to subdivide were part of the exchange. But their places are some distance from the Gray, which surely can't accommodate cattle from all over southwestern New Mexico. Pretty soon, a few folks from the northern group started talking about adapting the idea they had seen at work on the Gray's landscape to land nearer them. To create a mixed public-private rest-rotation unit of appropriate size, they'd need the Bureau of Land Management as a participant.

They would therefore have to create a conservation plan that moved beyond the ideas worked out on the Gray. But that is what the landscape in which they live and work calls for. If the northern ranchers get something going on their ground, still other ranchers may be interested. Some of them may modify the northern model to fit their situation—to deal with a slightly different landscape and the needs and concerns of yet a different set of landowners and managers.

We need a lot of such interactions. We need the ideas they will generate. We need to be developing a new ethic to guide our thinking about the economy and the environment. And ultimately, we need large businesses, especially resource-using businesses, as well as individual landowners and small businesses to understand and apply that ethic. Finally, to facilitate the needed transition, we also need consumers to expand that seminal willingness to pay more for an ecologically sound product until that willingness consciously motivates a broad transformation of the way we think about our purchases and the businesses we patronize.

A couple of years ago, the Conservancy welcomed a call from a large timber company. The object of the contact was to explore the possibilities for cooperation in conserving the biodiversity on the company's considerable private holdings. The company had just completed a transition in senior management, and there was a new urgency about improving environmental performance. Eventually, a senior vice president of the company asked the Conservancy to identify those corporate lands that were most important to biodiversity. Soon, TNC had checked the company's holdings against the Natural Heritage Program's network of biological databases (described in chapter 3) of the states in which the holdings were located and provided a working answer to the senior vice president's question. Representatives of the company responded by working with Conservancy representatives in a spirit of persistence and generosity to find ways to manage the sites that had been identified with due regard for their biodiversity values. The company and the Conservancy gradually forged an important agreement on a large and fairly high profile corporate forest. The agreement provides for a Conservancy role on certain management issues relating to that forest and includes a commitment by the company to place a conservation easement on a sizable part of it. The attempt at finding a way to cooperate therefore bore important fruit, and work will go on.

The success was real, and yet it left me full of questions as well as satisfaction. During the negotiations, I had begun to understand the way this company, and probably every large timber company, looks at things. A lumber mill is a substantial capital investment and an important place of work for company people. It is a possession to defend and protect. When a timber company owns a mill, a consideration that overrides almost any other is ensuring that a supply of timber that can economically be brought to it is available. The decisions the Conservancy's prospective partner found hardest to make always had to do with deciding to foreclose the possibility of harvesting all the timber at a site that met their specifications whenever demand called for it.

That, at base, is because at a level deeper than operational routine, success for that company is cutting, processing, and selling more timber. The company defines itself as I first described it—a timber company. Not that management wouldn't be reasonably happy to sell a little less lumber in year two than in year one at a little bigger margin and a little higher profit, but then their objective in year three would be to sell more timber at that bigger margin, or much more timber at a lesser margin so that their profit would exceed that of the previous year. They conceive of their business as selling lumber; success is selling more lumber. They have talked, I would bet, of being in the "building supply" business and perhaps the "forest management" business. But the talk hasn't changed the way they think. Not yet.

Another example: the Conservancy has a board member who also serves on the board of a well-known oil company. He called me one day and complained in the course of that call about the company's decision to abandon its solar energy business. The managers at the company wanted out of solar, though they had made it profitable, because the return on the equity they had invested in that renewable energy business wasn't up to fossil fuel standards; the growth potential was not big enough soon enough.

The timber company and the oil concern dwarf in dollars, staff, and influence the combined efforts of all the agencies, public and private, working on land conservation. By virtue of their economic power, they and their competitors large and small will have as profound an influence on the future of the natural world as any force in society. Right now, the central and inescapable operating facts of life for these companies have to do with the extraction, processing, and sale of resources—and the pursuit of those operations with the greatest vigor possible. Public questions about that approach, to the extent they are even understood, are offensive and threatening to such businesses. Private questions generate discomfort at best and, not infrequently, result in the termination of discussion in a fast-rising tide of mistrust. Such questions, after all, go to the soul of the business, and the implications are profound. As far as the rest of us are concerned, the central and inescapable fact of life about the way these businesses are run is that extracting and selling as fast as possible will ultimately damn us all.

The simple truths that shape the soul of most modern businesses day by day, quarter by quarter, and year by year—maximize profits, grow bigger, sell more—are leading us away from the real truth, which is more complex and more elusive. We want and need many of the products natural resource–using businesses supply. But we need them to be produced and used at less cost to the net. The demands we are making on the net are becoming too great—too many strands, too many webs will be weakened unless we find ways to meet our needs without extracting and selling and buying as fast as we can.

The failure to do so won't keep us from "saving" special areas, discrete systems, last great places. But the threads that hold together the ecological processes of the special areas, the systems, the landscapes, the hemisphere, and the globe are made of lands and waters that are not now, and will not be, for the most part, preserved in parks and protected areas. And as by hard use we wear thin those threads, two kinds of site on the net will become relatively heavy for the strands that support them. The intensively used lands, already burdensome for the net, will become even more so, and as the environment fails, the economy, which ultimately depends on the environment, will fail. What is less obvious at first glance is that the natural areas, the special places, will also become relatively heavy. As the supporting landscape threads that extend beyond their borders begin to break, the natural areas attached to those threads will become unbalanced and will also be doomed to fall eventually into the abyss. We can lose it all.

Conservationists can avoid losses for a time at the places we can afford to attempt to substitute by our resources for some of those supporting threads, but if such interventive management becomes universally essential to avoid losses, we will have acceded to the conversion of natural areas to wild botanic gardens and natural zoos. As mentioned in chapter 4, managing a zoo consumes enormous resources and requires constant attention. It's a last-ditch strategy for avoiding the worst losses of biodiversity, not a way to achieve harmony with the natural processes of our world. That harmony won't be easy to achieve, and we are only beginning to be familiar enough with the ecological arguments set forth earlier in this chapter, which, taken together, make it clear that we must as a society strive for it. We are only beginning to understand that regulating pollution and setting aside a few preserves or a few percentage points of the landscape will not be enough.

I remember being asked in the early 1980s how big a woods had to be to merit protection as a woods—to be a viable natural community—and answering that if it were any smaller than twenty acres, the biologists we worked with might admire it but ecologically it was a grove of trees. Subsequent research has cast doubt on the natural integrity of woods ten times the minimum size I named then. Effective minimum system size depends on what system you are trying to preserve, but asked the same question today, not only would I not say twenty acres, I'd be inclined to say that it was the wrong question.

The right question is not whether we can or ought to reach out and preserve a really nice twenty-acre woods but how we can hope to keep enough reasonably nice forest (not tree plantation—forest) on the continent, in the hemisphere, and on the planet. I still find my spirit refreshed in a small patch of good woods, but as a conservationist, if I can't find or afford a very large undisturbed woods, what I want to find is a magnificent twenty-acre grove of old-growth trees in the midst of an unimpressive but native forest of reasonable structural integrity that is contiguous over a thousand acres. If it were part of a patchy but partly forested landscape that covered a couple hundred thousand acres, I would put my resources into the relative conservation of that landscape before I would try to make a nature preserve of even a couple hundred biologically isolated acres of midwestern old growth. Acquisition of the twenty acres as the core of a relatively conserved but much larger area is a better long-term investment, and the contribution of the whole to the function of the hemisphere is much more likely to be important.

That conclusion brings us full circle in a sense. The discussion in this chapter is about the connections between system conservation and conservation in a larger context, a global context. But a different definition of system, a liberal analysis of the forest as a system, would have led me to the same conclusion reached earlier had I been using the five-S planning methodology when I was asked the question about how big a forest had to be. Good definitions of nearly any system we're working on will lead us to conclusions that call for something more than green measles: relative conservation—green shading—of big chunks of the seminatural landscape.

We can say which species and communities in the United States are most in need of conservation action with a fair degree of confidence. We know a lot less about the way the natural world *works* at any scale, and as the landscape unit about which we are asking gets bigger, we know yet less. But the research that is being done makes it clearer and clearer that at hemispheric and world scales, at least, Muir was right—everything *is* connected. I have argued for ignoring the implications of that worldview in planning for the conservation of the system, but they cannot be ignored in planning for the interaction of a series of system projects with the world's economies and governments such that the reasonable ecological function for the planet is secured.

The timber company mentioned earlier didn't ask the Conservancy to help it plan the relative conservation of corporate holdings that were not obviously and immediately significant habitat for rare elements of biodiversity. But that kind of planning must somehow be done, because much of such land falls into that critical 55 percent category—the relatively natural landscape in which many of the critical threads that bind together the ecological whole are and can be anchored.

It may be that to truly effect conservation of that kind of land, we need an external reorganization of the kind that corporations are so fond of managing internally. We may need a broadly based effort to reorganize twentieth-century capitalism and replace it with a form and structure that fit the needs of the twenty-first century, an operating economic philosophy that might be called postexploitive capitalism. Whatever the system is called, the changes required will happen quickest and the transition to them will be smoothest if they are driven both from within companies, through forward-looking management, and from without, by consumers who care and show it in the way they buy.

Natural resource–dependent businesses are already undergoing a change in mind-set, as is evidenced by the story of the timber company that approached the Conservancy for advice on dealing with biodiversity issues. But another leap forward is required if companies with large landholdings are going to play leading roles for society in planning to conserve a seminatural landscape that will provide ecologically useful support to both intensively used areas and natural areas. TNC has in the past been able to attract the support and participation of such businesses in part because of the narrow definition of its mission and the relatively small amount of land that mission seemed once to imply. Supporting the Conservancy's rare species conservation agenda is no longer avant-garde in the corporate world, but there are certainly plenty of businesses that aren't yet ready to do it. How many fewer businesses are prepared to meet the broader social responsibility of helping to preserve the integrity of the net? How many businesses will help define postexploitive capitalism? TNC's old approach was comfortable, but it is clear that TNC's mission and, more broadly, conservation of the net, requires more. We know, to put the point forward in a different way, that to preserve a species in nature, we have to look to its habitat and the ecological processes that sustain it. To preserve a natural community, we have to look to the surrounding landscape. Can we preserve a reasonable set of interacting species and natural communities without looking to the health of the landscape as a whole?

Conserving the health of the landscape as a whole is a task of nearly unimaginable size. System conservationists alone can probably do no better than start in the most important places, define system liberally at those places, monitor the status of the whole with existing measures, work at developing better measures so we are as well informed as we can be as we choose the next generation of conservation sites, and work hard at disseminating the lessons we learn about compatible economic use to the owners of the relatively natural parts of the landscape section. Companies that own the seminatural landscape must be open to that learning and must generate learning of their own. Society more broadly must achieve some awareness of lands and resources as it produces and consumes.

How do we make that leap? One thing we can do is expand by hundreds and hundreds the place-based, community-centered efforts to simultaneously deal with economic and environmental concerns. Such efforts have already begun in a few dozen places. From each of those projects, we have to promote extensions like those discussed with respect to the Clinch, the Gray Ranch, and Virginia's Eastern Shore. It has long been said that all politics is local. Ultimately, conservation is local too. Communities can gain control of their destinies if they make doing so a conscious goal and if they can gain access to ideas and tools that not only facilitate that process but also have special importance beyond that—they make it seem possible.

One group of farmers with whom the Conservancy is working, when asked what they wanted the future of their land to be, said they would like to keep farming, would like to see their kids have a chance to farm. When asked what they thought would actually happen, they said the land would be sold for development. It was their own land they were talking about—the decisions about sale were theirs to make. But at that point they felt as if loss of the land to farming—sale for development—was inevitable, that the future of the landscape they owned was not theirs to control. A good plan and a good planning process can change that. That's even true, at least up to a point, in communities that don't own some of the land that is critical to the maintenance of local ecological processes, in which perhaps timber or oil companies have big interests. Resource-using companies ultimately need the support of the communities in which they work, just as conservationists do. Imagine awakening like Rip Van Winkle after a draught of Hudson's brew to find as many well-trained directors of sustainable development in county offices across the nation as we now see directors of tourism or offices of economic development—if communities decide that is what they want their local government to do, it isn't an unattainable vision. Northampton County, one of the poorest in Virginia, hired a director of sustainable development in 1994.

A second way to change the relationship between businesses and the environment—to expand on a point already made—is to change the decisions each of us makes about what we will buy and what we require of the supplier of our needs. Through our buying decisions, we can stiffen the resolve of companies to be leaders in defining postexploitive capitalism. Collectively, we make a market—for some retail companies, *the* market. If we demand lumber that has been produced as sustainably as humankind knows how, timber companies will supply it. If we stop buying wood not certified to our standards, they'll pretty quickly supply wood that is. Nearly 70 percent of respondents in a recent poll, for example, said they would be willing to pay more for furniture made from wood taken from a sustainably managed forest.

If even some of us (there are so many of us now that 10 percent is a big, big market) showed a marked preference for vehicles powered by electricity, and electricity generated from sustainable sources, that is what car companies would supply, and that is what electric companies would transmit. When asked why they are not marketing electric cars, the major auto companies sometimes answer that there are a lot of engineering problems to be solved before they can produce cars that satisfy consumers' demands. But the real problem is that demand isn't intense enough. With sufficient demand, cited problems would become the focus of competition in the industry and would quickly be overcome. The most intransigent of companies will change its practices and maybe even examine its soul and transform its self-image, when the alternative is to be left behind by the market. And if it doesn't, it will be left behind.

Obviously, substantial capital is required to enter many businesses—there are, as economists say, barriers to entry—and without ready competitors, the preferences of the market will be slowly and imperfectly reflected. According to analyst and commentator Jessica Matthews, even electric cars, as demanding of capital as they must be to produce, can be competitively manufactured by companies that aren't of Fortune 500 size. If public demand for reduced pollution is sustained and government takes sufficient note of that demand and maintains air pollution standards, our multilayered national automobile industry may have to start changing the ways cars are made, maintained, serviced, and marketed—or be left behind.

In *The Ecology of Commerce,* Paul Hawken blames corporations for steering us in the wrong direction through their offerings and their messages, but as a society we've rewarded corporations for behaving in that way by responding as their advertising executives predicted we would. I find more hope in the reservoir of power he cites in his final chapter: our underdeveloped potential as consumers to make the right choices. Failure to insist on the power of that potential allows us to deny that we have the responsibility of exercising it: if it is corporations that are destroying the environment, then the environment can be saved by bringing corporations into line—and that is somebody else's responsibility. But if corporations aren't us (and all corporations *are* some of us), they will respond to us; we don't see many cars with fins and chrome any more. We're not powerless, and responsibility for the environment is as much ours—through the market signals we send—as it is that of the corporations, which, after all, supply our demand and do not have irresistible power to create it. Even corporations that are alert and agile do not usually come equipped with a conscience and will develop one only if we insist through our buying choices that they do so.

The tradition among environmentalists is to look to the government as the source of power to save the environment. Government can be, has been, and must be such a force. All of the will, however, we have summoned from it thus far hasn't been enough, and for the moment, the steady movement of government policy in this country toward environmental protection has slowed.

There is only one other force with the potential of government. It is the other vehicle for the expression of collective will, and in the end, its power far exceeds that of government. It is the market, and we who care about the natural world have to learn how to keep it from running wild over wilderness and harness it to bring about change that saves. With even a little success, we might provide Jose Lutzenberger with the text for a new sermon.

CHAPTER 14

There Are Tigers in the World

A few days after our youngest child, our daughter, was born, my family gathered around the supper table to celebrate the addition. We decided to take a moment before we ate to think about the promise inherent in our newly expanded family. My wife, Mary, spoke first, saying how glad she was to be back home with us. She spoke warmly, too, about how beautiful Anna was and how wonderful a part of our lives she would be. My older son, Joseph, followed in the same mode, and so did I. John, then four, had listened carefully and appreciatively as he waited for his turn.

"I'm thankful," he said, "that there are tigers in the world."

We all smiled. The words were right for a moment in which we were celebrating life, and they swelled the wave of warmth that filled the room. I think about what John said pretty often. He knew that some animals were disappearing from the earth, as the dinosaurs had long ago. He loved tigers. He had a four year old's fascination with their color, speed, and grace and a well-protected four-year-old boy's admiration for their ferocity. And I think that in that circle of parent and child, John somehow understood that the prayer for tigers was properly made by his generation in the presence of mine.

An amazing thing has happened to humanity. Almost overnight, the promise of Genesis has been realized. We have dominion over the earth. We've come to the point at which a salmon fisherman working out of Alaska could say, "We're lucky to live in a corner of the world where we're still vulnerable to nature." Lucky! Vulnerable. That surely used to be the way it was. "I'm thankful there are tigers in the world."

Our preparation as a species for our new status as monarchs is suspect. We put in several hundred thousand years of living with nature as a provider, but a frequently moody and sometimes hostile one; we often had to protect ourselves from nature so that we could survive. Then followed six or seven thousand years in which we domesticated the nature nearest us to make living a little safer, a little easier, and a little more predictable. Then

Figure 14-1. *Panthera tigris.* Howard B. Quigley/HWI. Save The Tiger Fund

three or four hundred years of extracting wealth from nature so that some of us could live pretty well. That brings us near the present—during the last couple of decades we have seen the dawning realization that we have dominion and that the principal issue of our existence as a species has been turned on its head: that now we need to protect nature from *us,* so that it, and we, can survive.

The realization of the biblical promise of dominion—"Let them have dominion over . . . every living thing"—was a long time coming, but for all that, we seem ill prepared to exercise it. The text from Genesis itself is noteworthy for the shallow reading it has received from environmentalists and exploiters alike. The King James version does use the word "subdue," but it also says that we who were to have dominion were made in God's likeness and that God, after creating the beasts and birds and trees and herbs, is said to have thought that the life that preceded man was good.

Only a little less obvious are the implications of dominion. A monarch's security and wealth reside in the prosperity of his or her subjects. A ruling dynasty that impoverishes the realm is ultimately poor. Are we able to destroy nature as we know it? We can. Is it our destiny, our fate, or our divine role to do so? Of course not. We can be wise and benevolent rulers. We can look over creation and see that it is good. We can treat well what is good—every beast, fowl, and shrub.

Our dominion is oddly unbalanced. The clearest evidence that it exists is our destructive capability. We cannot create nature; we can only live with it or push it out of our way and try to live without it. Toward the end of our domestication era and during our years of exploitation, we developed a sense of separation from nature. As we glorified this supposed transcendence, disagreeable things became "beastly," and in intellectual circles, nature became a state that was, as Thomas Hobbes said, "nasty, brutish, and short."

This was a harmless enough delusion until our numbers and our technology made one of a number of possible futures the end of nature as we know it. Now we have to reestablish a broad understanding that, as Joseph Campbell has said, we come "out of the earth, rather than having been thrown in here from somewhere else," and that indeed "we are the consciousness of the earth."

That insight implies two linked and opposing realities, the yin and yang of our relationship with nature. First, we are from and part of a biological community to whose laws we are in the end, subject. As Denis Hayes, president of the Bullitt Foundation, has said, "You can't break nature's laws; you can only prove them." Second, on earth, it is we and we alone who can observe, reflect upon, and make judgments about the course of events that affect the earth. Unlike any other part of the biological community, we can effect change, and thus we inevitably are charged with responsibility for changing our behavior if, conscious of its effect on earth, on us, we find it appropriate to do so. We are subject to nature's laws, but we are not players in a drama the script for which is beyond our control. We can decide what kind of citizens we'll be in our biological community.

It may be that my sense that our instinct is to be good, loving citizens is wrong. It may be that nature the unpredictable, dangerous force is the dominant residual memory that we have. But in my experience, you don't have to teach children to love nature.

My son Joe studied the blossoms of newly opened wildflowers intently every spring from the age of one and a half. When John could barely talk, he was sitting one morning on the kitchen counter while I was getting breakfast together, and he pointed out the window toward the bird feeder, burbling some multisyllabic phrase. I affirmed the presence of a "bird" out there. Two more repetitions; I could hear five syllables. Two more, and I made them out. Not bird, "white-throated sparrow." And so it was.

Every cecropia cocoon we have hatched has brought all the neighborhood kids within hailing range over to watch, uncharacteristically still and attentive, almost reverent. Our kids, always pushing the bus deadline in the morning during the school year, were awakened with a soft word, before the light, to go over to the beaver lodge we discovered one year on vacation. Joseph has a memory that will be his for life of a bear cub that he saw first, from a canoe at daybreak, at a place called Turtle Rock, upstream from an Adirondacks lake.

Some of us don't grow completely out of that fascination with the natural world. I still feel a tangible sense of refreshment upon a walk through any kind of a woods beyond the audible range of traffic. Those attenuated symbols of nature mentioned in chapter 10—the duck prints on the wall, the tropical office plants—are there because even those who seem to have left nature behind haven't completely escaped it.

It may be that some are immune, born without openness to the wonder and beauty of the natural world. But I think more of us just accept what we take to be learning—that other things are more important. We accept as part of growing up the replacement of our love for nature with knowledge not of the natural world but of humankind's world within a world. Agriculture, automobiles, fashion, finance, computers, collector's items, weed whackers, and war. The recent Gulf War was incredible for many reasons, not least the environmental destruction purposely inflicted by Iraqi military forces. It was noted at the time that the rules of war once proscribed acts meant to destroy the capacity of the land to sustain life. No more. We've "grown up" about that, too.

Perhaps the problem is that humanity is like an adolescent in relation to the technologies of the industrial and postindustrial world: a lot of physical capability and bright as can be, but incapable of wise behavior. It seems clear, in fact, that we suppress some of the unconscious wisdom with which we were born. We would do well to unlearn the order of priorities we assumed was part of growing up and listen to the messages, of which we alone can make sense, that come to us as the consciousness of the earth.

African conservationist Perez Olindo says that to do conservation, we need only facilitate a process. He says that people need to touch nature and test it. If they can do that, they will, he says, understand it, and like it. "If you like something," he concludes, "you will protect it." Touch, test, understand, like, protect: a poet's path, a way to Campbell's consciousness.

But many of us live insulated lives in which nature is somewhere else or is brought in, carefully groomed, a little at a time. That consciousness of the earth that is essential to our future is elusive. What are we to make of the environmentalists' refrain that our economic health ultimately depends on ecological health when meat comes from refrigerated display cases, bread from the shelf, water from the tap, the stuff our homes are made of and filled with from who-knows-where/hadn't-thought-about-it? We can be deluded, as many have been, into thinking that we no longer need nature, that we can substitute our production systems for anything and everything nature provides us. Some who are environmental advocates have suggested that we stop trying to make a case that we need nature for anything except the joy it brings us. That view denies us even the essential chance to reclaim a connection with things greater and more wondrous than we are—because it denies that nature is greater and more wonderful than we are.

But more important, that view ignores a long line of technological fixes that weren't, which prove we never see the whole picture, and illustrate that ecological systems really are more complex than we can think.

Visit the reconstruction of a Mayan city and ask yourself whether the achievements of the civilization whose shadow you are seeing do not rival anything modern culture has produced. And then reflect on the likelihood that an ecological collapse occasioned the disappearance of the social organization that generated those monumental plazas, those decorated walls, that government, that priesthood. The Aswan Dam is a fine thing, surely, but we haven't found the technological fix for the sardine fishery that collapsed when the run of the Nile was fundamentally altered.

We simply do not think at the depth required to manipulate nature

Figure 14-2. Joseph, Bill, and John Weeks clamming, Hog Island, Virginia. © John M. Hall

without unexpected consequences. We had the technology to introduce the Nile perch into Lake Victoria in central East Africa to improve the sport fishery there. People living around the lake had traditionally eaten tilapia, a relatively small native fish. They preserved their catch by drying it in the sun. The perch, however, also found the tilapia to be good eating. The tilapia escaped at first by swimming into the deep water, where they found refuge from the sight-feeding perch. The perch then turned to other fish. Those fish fed on algae, and when their numbers were sufficiently diminished, the algae grew thick. As the thick mass of algae died, it sank in greater amounts than ever before to the depths of the lake, using up oxygen as it decayed. As oxygen became scarce in the depths of the lake, the tilapia were forced back up into the jaws of the perch. The lakeside villagers adjusted, as one might expect, and began to eat perch. But sun-drying did not work on the oily perch; it had to be smoked. The source of the wood fuel for smoking was the forest that surrounded the lake, and as the forest was depleted, the result was soil erosion and further degradation of the lake—another fix that wasn't.

You can't break nature's laws; you can only prove them. To pursue Hayes's metaphor a bit, nature's laws are generous. They allow for a lot of bending. But that is a little scary because in the end, Hayes is right: bend, bend, bend . . . proof.

At the time of proof, consciousness of our negligent performance as the eyes of the earth becomes painfully acute. At its best, conservation is about avoiding that moment.

I work in conservation and find energy in and for it because in spite of lots of reasons for despair, I think we have it within us not only to avoid an awful confrontation but to convey to our children a world in which development and preservation are balanced well short of the crisis point.

When The Nature Conservancy and other conservation organizations have sown the seeds of ideas that make it practical for people to live and make a living with nature, the seeds have fallen on fertile ground. Something deep inside many of us tells us we're better off, more secure, safer, and happier if the ecosystems in which we live and work are sound and healthy.

Tigers still persist in a few places where there are ecosystems still big enough and sound enough to support the rarefied top of the trophic pyramid. And I'm with John. I'm glad there are still tigers in the world. We ought to judge our success as a generation of stewards of the planet in part by whether we keep them.

Notes and Comments

Chapter 3

The book mentioned at the beginning of the chapter, Phillip M. Hoose, *Building an Ark: Tools for the Preservation of Natural Diversity Through Land Protection* (Covelo, Calif.: Island Press, 1981), is an exceptionally rich source of information for land conservationists. Though it is out of print, copies should be available in most good libraries. I have quoted from Hoose's introduction at pages xv and xiv.

Figures on the amount of land converted each year to urban or suburban use are provided in the summer 1991 issue of the newsletter of the National Growth Management Leadership Project. The lead article asserts that "between 1.2 and 1.5 million acres of rural land are converted to development and urban use" each year (*Developments* 2, no. 1 [summer 1991]: 2). Stephen Meyer, in *The New Republic*, no. 4 (15 August 1994): 152, states: "Human development consumes more than 200 acres every hour."

The statistics on the amount of earth moved each year are from Richard Monastersky, "Earthmovers," *Science News*, 146, nos. 26, 27 (December 1994): 432.

Estimates of the number of species driven to extinction in some unit of time vary widely among current sources. I have presented a more conservative figure than many analyses would yield.

Chapter 4

The discussion of the effect of ozone in the atmosphere is adapted from the excellent discussion of the same topic in Donella H. Meadows, Dennis L. Meadows, and Jorgen Randers, *Beyond the Limits* (Post Mills, Vt.: Chelsea Green, 1992).

Mention of the loss of arable land due to excessive concentration of salts is in Albert Gore, *Earth in the Balance: Ecology and the Human Spirit* (Boston: Houghton Mifflin, 1992), 125.

Statistics on the relative endangerment of freshwater fauna were assembled by The Nature Conservancy's Larry Master and his colleagues, relying to a significant degree on data assembled by members of the Natural Heritage Program's network of biological databases.

The numbers on housing foundation excavation and grading are from Richard Monastersky, "Earthmovers," *Science News* 146, nos. 26, 27 (December 1994): 432.

A comparison of statistics from two agencies of the relatively flat state of Indiana reveals that three times as much soil (by weight) is lost every year through erosion, as the combined annual weight of the three leading crops produced. A comparison of such statistics in other states would be expected to yield a result at least as dismaying.

The estimate that some 40 percent of our prescription medicines utilize extracts from wild organisms is from Edward O. Wilson, *The Diversity of Life* (Cambridge, Mass.: Harvard University Press, 1992), 283, 285.

Recent research on the impact of birds on trees is reported in Carol Kaesuk Yoon, "Eating Like a Bird," *New York Times,* 8 November 1994, p. B11.

The Ehrlich and Wilson article suggesting a moratorium on the development of undeveloped land in the United States is "Biodiversity Studies: Science and Policy," *Science* 253, no. 5021, (16 August 1991).

Chapter 5

I credit the useful clarification represented by the phrase "an ecosystem approach to conservation" rather than "ecosystem conservation," to The Nature Conservancy's director of science, Deborah Jensen.

Chapter 6

The John Muir quote appears in Stephen R. Fox, *The American Conservation Movement: John Muir and His Legacy* (Madison: University of Wisconsin Press, 1981), 291.

Chapter 8

One situation in which population has an important local effect is where there is excessive demand for local natural resources. Marshall Murphree, a principal in the establishment of Zimbabwe's widely praised CAMPFIRE program of community participation in management of natural resources, says that a community must be able to define and control membership (implying an ability to limit in-migration) if it is to serve as a base for real and sustainable conservation. That ideal is difficult, in the age of the mobile and global village, to achieve. Yet Murphree believes that community-based conservation is the only conservation that will be effective in the end. Local population pressure, particularly from in-migration, perhaps should be recognized as a source of stress in some system conservation projects.

Chapter 10

At the conclusion of the discussion of considerations site conservationists should employ as they decide which kinds of community problems to address, I assert that the final decision belongs (constrained, of course, by the values of the organization) in the field. The allocation of such decisions could itself be the subject of a book—and it more or less has been. I recommend Peter Block, *Stewardship: Choosing Service over Self-interest* (San Francisco: Berrett-Koehler, 1993).

When I discuss with conservationists strategies for compatible economic development, I am often asked whether I do not worry that enhanced economic opportunity will attract people to the site we are trying to conserve and create a whole new round of problems. Generally, I would rather deal with that kind of success than leave people with no alternative but to overexploit the environment in order to make a living. A substantive response can emerge only with thinking and planning about the particular conservation site. Such advance planning will be especially important at international conservation sites with mobile populations so devoid of economic opportunity that even a small success might create significant pressure from in-migration.

A World Wide Web site devoted to the emerging field of "conservation-based development" is being developed under the sponsorship of the Ford Foundation. Many of the themes discussed here in the context of compatible development will be explored in that electronic forum.

The statistics on past membership of conservation organizations are from Stephen R. Fox, *The American Conservation Movement: John Muir and His Legacy,* (Madison: University of Wisconsin Press, 1981).

The quotation from the spokesman for the South Florida Water Management District is from the *Washington Post,* 17 January 1995, p. A3.

Chapter 11

The lodgepole pine–hemlock–larch community life cycle example is from Peter J. Cattelino, Ian R. Noble, Ralph O. Slayter, and Stephen R. Kessell, "Predicting the Multiple Pathways of Plant Succession," *Environmental Management* 3, no. 1 (1979):41-50.

For some guidance on where to look for social and economic data, see Priscilla Salant, and Anita J. Waller, *Guide to Rural Data,* rev. ed. (Washington, D.C., and Covelo, Calif.: Island Press, 1995). Useful guidance is also available in Ray Rasker, Jerry Johnson, and Vicky York, *Measuring Change in Rural Communities: A Workbook for Determining Demographic, Economic, and Fiscal Trends* (Washington, D.C.: The Wilderness Society, 1994).

Chapter 12

The comment about the first hundred years of "taking" jurisprudence under the U.S. Constitution is from Bosselman, Callies, and Banta, "The Taking Issue" (1973), as quoted in Jacob A. Beusher, Robert R. Wright, and Morton Gitelman, *Land Use,* 2nd ed. (St. Paul, Minn.: West, 1976).

The points attributed to Celinda Lake were made in an address to a conference in Santa Fe, New Mexico, in June 1995.

Chapter 13

There isn't a venerable legal tradition to refer to in support of plant conservation. Perhaps it seemed natural enough to deny landowners' claims of wild animal ownership because of the animals' ability to move around. Probably because plants don't move around freely, it hasn't commonly been asserted that wild plants are among the things over which an owner of real property cannot exert unfettered control. Wild plants surely deserve conservation action no less vigorous than that afforded wild animals. We ought to consider whether a case can be made that for legal purposes, the closest analog in the plant world to an individual animal is not an individual plant but a reproducing population of plants with associated species such as pollinators. If that construct is sound, there is a foundation for regulating activities that affect plants that is analogous to the foundation on which animal regulation stands.

The statistics on the percentage of the U.S. landscape that is in conservation management are from Reed F. Noss, and Allen Y. Cooperrider, *Saving Nature's Legacy: Protecting and Restoring Biodiversity* (Washington, D.C., and Covelo, Calif.: Island Press, 1994).

The figures for the portions of the U.S. landscape that are not in conservation management and are not intensively used are from Dale Curtis and Barry Walden Nash (eds.), "Environmental Quality: The Twenty-Second Annual Report of the Council on the Environmental Quality Together with the President's Message to Congress (Washington, D.C.: Council on Environmental Quality, 1992).

The discussion of the amount of conserved or relatively conserved land required to sustain an acceptable amount of biodiversity makes mention of work that has been done on island biogeography. Island biogeography theory, as explained briefly in Edward O. Wilson, *The Diversity of Life* (Cambridge, Mass.: Harvard University Press, 1992), generally holds that with a tenfold loss of habitat, overall species diversity will be diminished by about half. The relationship surely varies, with type of habitat and nature of species involved, and some scientists question the applicability of results obtained on islands surrounded by ocean to islands surrounded by land. As a boundary, land will undoubtedly behave differently from water, but the trend described in the theory will undoubtedly find expression.

See Carol Kaesuk Yoon, "Eating Like a Bird," *New York Times*, 8 November 1994, p. B11, regarding the assertion that trees need birds.

With respect to whether there are alternatives to conservation or relative conservation, it seems likely that we can reinforce many natural systems and, ultimately, the whole—what I have called the "net"—by conserving corridors that link natural nodes in the net. Those corridors must be designed to account for potential ecological problems such as predation rates within the corridors and linkages that serve as pathways for transmission of disease. But because the corridor concept has such potential utility in a landscape that must accommodate many nonnatural uses, we are probably better advised to work at minimizing the effect of problems instead of abandoning corridors as a possible conservation tool. Even a very well developed system of corridors between natural areas would not substitute, though, for the reasonable ecological functioning of the 55 percent of the landscape not devoted to intensive uses of parks and preserves—such a system would not make it possible to treat the seminatural landscape as if its contribution to ecological function did not matter.

Consumers' willingness to pay more for furniture made from sustainably harvested wood is documented in Dawn Winterhalter and Daniel Cassens, "United States Hardwood Forests: Consumer Perceptions and Willingness to Pay" (Purdue University, Department of Forestry, August 1993). Similar evidence relating to T-shirts can be found in Laura M. Litvan, "Going Green in the '90s," *Nation's Business,* February 1995, p. 32. Also, in a late 1995 telephone poll conducted by Market Revelations, Inc., for The Nature Conservancy, consumers expressed a strong interest in and a willingness to pay a substantial premium for "conservation beef."

Chapter 14

The story of the Nile perch is summarized in the *Washington Post,* 5 June 1989, p. 3.

Index

About the Author

W. William Weeks is director of the Center for Compatible Economic Development. An attorney, his conservation experience includes six years of service as director of The Nature Conservancy's Indian program and six years as the Conservancy's chief operating officer. The Center works with communities to develop businesses, products, and land uses that conserve ecosystems, enhance local economies, and achieve community goals. It is an operating unit of The Nature Conservancy, of which the author is currently vice president. Loudoun County, Virginia, is home for both the author and the Center.